日経サイエンスで鍛える科学英語
医療・健康編

Scientific English with
SCIENTIFIC AMERICAN®/
NIKKEI SCIENCE

日経サイエンス編集部 [編]

発行 日経サイエンス社

Scientific American trademarks used with permission of Scientific American, Inc.

まえがき

　本書は月刊科学雑誌『日経サイエンス』に翻訳掲載した SCIENTIFIC AMERICAN 誌の健康科学コラム「The Science of Health」(日経サイエンスでの連載名は「ヘルス・トピックス」) の主要記事を原文と対照して読めるようにしたものであり,『日経サイエンスで鍛える科学英語』(2011 年刊) および『日経サイエンスで鍛える科学英語 2 [読解編]』(2013 年刊),『日経サイエンスで鍛える科学英語　ノーベル賞科学者編』(2015 年刊) の続編にあたる。
　「興味深い科学ストーリーを英語で読むことによって, 楽しみながら科学英語に親しむ」という一石二鳥の狙いはこれまでと同じだが, 既刊の 3 冊が長文記事の抜粋であったのに対し, 本書では各コラムの全文を収載した。各記事の長さは容易に通読できる分量 (雑誌では見開き 2 ページ) であり, 起承転結を含めた文章の構成を把握できる。著者は米国で活躍する気鋭の科学ライターやジャーナリストで, その明快で歯切れの良い英文と, 時にウイットに富む表現に接することができるのも魅力といえるだろう。
　読者が頻繁に辞書をめくらずにすむよう, 英文の右側に語注「Vocabulary」を併記した。また, 知識として押さえておくと理解の助けになると思われる専門用語について簡単な説明「Technical Terms」を掲載した。
　一般に本誌「日経サイエンス」の訳文は, より読みやすくするため逐語訳にはしていない。原文にない内容を補足したり, 部分的に順序を入れ替えたりしている例があり, 特に本書に収載したコラムでは論旨を損なわない範囲で省略した部分もある。英語の試験で求められるような逐語訳ではないことをお断りしておく。
　なお, 本誌ウェブサイト (http://www.nikkei-science.com) には「英語で読む日経サイエンス」というコンテンツがある。各号に掲載した主要記事から毎月 1 ～ 2 本を選び, その冒頭から数パラグラフの原文・訳文を並べて表示したものだ。2005 年夏にスタートし, 2017 年 11 月時点で 220 本あまりをアップしてある。また誌上では 2009 年 5 月号から「今月の科学英語」を連載している。既刊書 3 冊を含め, これらも併せてご利用いただきたい。

2017 年 12 月　　　　　　　　　　　　　　　　　　日経サイエンス編集部

日経サイエンスで鍛える科学英語［医療・健康編］●目次

まえがき

1　肥満&生活習慣
That Craving for Dessert
快楽飢餓と闘う 6

Artificial Sweeteners Get a Gut Check
人工甘味料で肥満に？ 17

Can Brown Fat Defeat Obesity?
褐色脂肪で肥満を防げるか？ 27

Gut Reactions
腸内細菌と肥満 37

Is Fasting Good for You?
プチ断食は健康によい？ 48

2　脳&老化
Blue Light Blues
青色光が奪う眠り 58

A Pain in the Brain
片頭痛に予防薬 68

Brain Food
地中海食と脳の健康 78

A Turn for the Worse
回転性めまい 88

Can We Stop Aging?
老化を止められるか？ 97

3　潜むリスク

Cold Comfort
お寒い冷却療法 108

The Acupuncture Myth
鍼治療の神話 119

Overreaction
過剰反応のアレルギー検査 129

Deadly Drug Combinations
危うい薬の飲み合わせ 139

When DNA Means "Do Not Ask"
遺伝子検査のジレンマ 149

4　先端医療

Busting Blood Clots
危険な血栓を取り除く 158

Cancer Gene Tests Provide Few Answers
がん遺伝子検査のいま 167

A Surprising Fix for Sickle Cell
鎌状赤血球症に驚きの治療法 178

The Paradox of Precision Medicine
個別化医療の矛盾 189

The Not So Silent Epidemic
広がる睡眠時無呼吸症 199

凡例　※翻訳記事では，よりわかりやすく読みやすい文章にするため，内容を補強あるいは省略するなどの変更を加えている場合があり，必ずしも英文記事の逐語訳とはなっていない。また，改行位置の変更など日経サイエンスの誌面掲載時とは異なる文章になっている場合がある。
　　　※各記事の掲載時期を，表題の下にSCIENTIFIC AMERICANおよび日経サイエンスの掲載号で示した。

装丁：八十島博明
カバーイラスト：中村知史
DTP：GRID

3

肥満 & 生活習慣

That Craving for Dessert
快楽飢餓と闘う

Artificial Sweeteners Get a Gut Check
人工甘味料で肥満に？

Can Brown Fat Defeat Obesity?
褐色脂肪で肥満を防げるか？

Gut Reactions
腸内細菌と肥満

Is Fasting Good for You?
プチ断食は健康によい？

That Craving for Dessert
快楽飢餓と闘う

ジャンクフードは脳を乱して食欲を暴走させる

F. ジャブル（SCIENTIFIC AMERICAN 編集部）

掲載：SCIENTIFIC AMERICAN January 2016, 日経サイエンス 2016 年 7 月号

Matthew Brien has struggled with overeating for the past 20 years. At age 24, he stood at 5'10" and weighed a trim 135 pounds. Today he tips the scales at 230 pounds and finds it particularly difficult to resist bread, pasta, soda, cookies and ice cream—especially those dense pints stuffed with almonds and chocolate chunks. He has tried various weight-loss programs that limit food portions, but he can never keep it up for long. "It's almost subconscious," he says. "Dinner is done? Okay, I am going to have dessert. Maybe someone else can have just two scoops of ice cream, but I am going to have the whole damn [container]. I can't shut those feelings down."

Eating for the sake of pleasure, rather than survival, is nothing new. But only in the past several years have researchers come to understand deeply how certain foods—particularly fats and sweets—actually change brain chemistry in a way that drives some people to overconsume.

Scientists have a relatively new name for such cravings: hedonic hunger, a powerful desire for food in the absence of any need for it; the yearning we experience when our stomach is full but our brain is still ravenous. And a growing number of experts now argue that hedonic hunger is one of the primary contributors to surging

Vocabulary

overeating 食べ過ぎ
tip the scales at~ 重さがある（scale は「天秤」「重量計」）

portion 食事の一人前, 食事量
keep it up 続ける

scoop ひとすくいの量

survival 生存

overconsume 過食する

craving 渇望
hedonic hunger 快楽飢餓

yearning 切望
ravenous 飢えた, 腹ぺこの

surging 高まる, 押し寄せる

obesity rates in developed countries worldwide, particularly in the U.S., where scrumptious desserts and mouthwatering junk foods are cheap and plentiful.

"Shifting the focus to pleasure" is a new approach to understanding hunger and weight gain, says Michael Lowe, a clinical psychologist at Drexel University who coined the term "hedonic hunger" in 2007. "A lot of overeating, maybe all of the eating people do beyond their energy needs, is based on consuming some of our most palatable foods. And I think this approach has already had an influence on obesity treatment." Determining whether an individual's obesity arises primarily from emotional cravings as opposed to an innate flaw in the body's ability to burn up calories, Lowe says, helps doctors choose the most appropriate medications and behavioral interventions for treatment.

ANATOMY OF APPETITE

Traditionally researchers concerned with hunger and weight regulation have focused on so-called metabolic or homeostatic hunger, which is driven by physiological necessity and is most commonly identified with the rumblings of an empty stomach. When we start dipping into our stores of energy in the course of 24 hours or when we drop below our typical body weight, a complex network of hormones and neural pathways in the brain ramps up our feelings of hunger. When we eat our fill or put on excess pounds, the same hormonal system and brain circuits tend to stifle our appetite.

By the 1980s scientists had worked out the major hormones and neural connections responsible for metabolic hunger. They discovered that it is largely regulated by the hypothalamus, a region of the brain that contains nerve cells that both trigger the production of and are exquisitely sensitive to a suite of disparate hormones.

Vocabulary

obesity 肥満
scrumptious とてもおいしい
mouthwatering よだれの出そうな

coin 造語する

palatable 味のよい

innate flaw 内在的な欠陥

metabolic hunger 代謝的空腹
homeostatic ホメオスタシスによる

dip into 〜を使う

ramp up 高める
eat one's fill たらふく食べる

stifle 抑える

hypothalamus 視床下部

disparate 異なる, 異種の

As with so many biological mechanisms, these chemical signals form an interlocking web of checks and balances. Whenever people eat more calories than they immediately need, some of the excess is stored in fat cells found throughout the body. Once these cells begin to grow in size, they start churning out higher levels of a hormone called leptin, which travels through the blood to the brain, telling the hypothalamus to send out yet another flurry of hormones that reduce appetite and increase cellular activity to burn off the extra calories—bringing everything back into balance.

Similarly, whenever cells in the stomach and intestine detect the presence of food, they secrete various hormones, such as cholecystokinin and peptide YY, which work to suppress hunger either by journeying to the hypothalamus or by acting directly on the vagus nerve, a long, meandering bundle of nerve cells that link the brain, heart and gut. In contrast, ghrelin, a hormone released from the stomach when it is empty and blood glucose (sugar) levels are low, has the opposite effect on the hypothalamus, stimulating hunger.

By the late 1990s, however, brain-imaging studies and experiments with rodents began to reveal a second biological pathway—one that underlies the process of eating for pleasure. Many of the same hormones that operate in metabolic hunger appear to be involved in this second pathway, but the end result is activation of a completely different brain region, known as the reward circuit. This intricate web of neural ribbons has mostly been studied in the context of addictive drugs and, more recently, compulsive behaviors such as pathological gambling.

It turns out that extremely sweet or fatty foods captivate the brain's reward circuit in much the same way that cocaine and gambling do. For much of our evolutionary past, such calorie-dense foods were rare treats that would have provided much needed sustenance, especially in dire times. Back then, gorging on sweets and fats whenever they were available was a matter of survival. In contemporary society—replete with inexpensive, high-calorie grub—this instinct works against us. "For most of our history the challenge for human beings was getting enough to eat to avoid starvation," Lowe says, "but for many of us the modern world has replaced that with a very different challenge: avoiding eating more than we need so we don't gain weight."

Research has shown that the brain begins responding to fatty and sugary foods even before they enter our mouth. Merely seeing a desirable item excites the reward circuit. As soon as such a dish touches the tongue, taste buds send signals to various regions of the brain, which in turn responds by spewing the neurochemical dopamine. The result is an intense feeling of pleasure. Frequently overeating highly palatable foods saturates the brain with so much dopamine that it eventually adapts by desensitizing itself, reducing the number of cellular receptors that recognize and respond to the neurochemical. Consequently, the brains of overeaters demand a lot more sugar and fat to reach the same threshold of pleasure as they once experienced with smaller amounts of the foods. These people may, in fact, continue to overeat as a way of recapturing or even maintaining a sense of well-being.

Emerging evidence indicates that some hunger hormones that usually act on the hypothalamus also influence the reward circuit. In a series of studies between 2007 and 2011, researchers at the University of Gothenburg in Sweden demonstrated that the release of ghrelin (the hunger hormone) by the stomach directly increases the release of dopamine in the brain's reward circuit. The researchers also found that drugs that prevent ghrelin from binding to neurons in the first place curtail overeat-

Vocabulary

treats ごちそう
sustenance 食べもの, 栄養
dire 厳しい
gorge on つめこむ
replete with 〜でいっぱいの, 飽食した
grub 食べもの

taste bud 味蕾

spew どっと吐き出す
dopamine ドーパミン

saturate いっぱいにする, 飽和させる
desensitize 鈍感にする, 減感作する
receptor 受容体

threshold 閾値, 水準

recapture 再びとらえる, 呼び起こす

neuron ニューロン, 神経細胞
curtail 抑える, 減らす

ing in people who are obese.

Under normal conditions, leptin and insulin (which become abundant once extra calories are consumed) suppress the release of dopamine and reduce the sense of pleasure as a meal continues. But recent rodent studies suggest that the brain stops responding to these hormones as the amount of fatty tissue in the body increases. Thus, continued eating keeps the brain awash in dopamine even as the threshold for pleasure keeps going up.

CURBING CRAVINGS

A kind of surgery that some obese people already undergo to manage their weight underscores ghrelin's importance in weight control and has provided some of the biological insights into why many of us eat far beyond our physiological needs. Known as bariatric surgery, it is a last-resort treatment that dramatically shrinks the stomach, either by removing tissue or by squeezing the organ so tightly with a band that it cannot accommodate more than a couple of ounces of food at a time.

Within a month after such surgery, patients are typically less hungry overall and are no longer as attracted to foods high in sugar and fat—possibly because of changes in the amount of hormones that their much smaller stomach can now produce. Recent brain-scanning studies reveal that these reduced cravings mirror changes in neural circuitry: postsurgery, the brain's reward circuit responds much more weakly to the images and spoken names of tempting foods, such as chocolate brownies, and becomes resensitized to smaller amounts of dopamine.

"The idea is that by changing the anatomy of the gut we are changing levels of gut hormones that eventually get to the brain," says Kimberley Steele, a surgeon at the Johns Hopkins University School of Medicine. A few studies have documented lower levels of hunger-stimulat-

Vocabulary

insulin インスリン

fatty tissue 脂肪組織
awash あふれて

underscore 裏づける

bariatric surgery 肥満外科手術, 減量手術
last-resort 最後の手段の

accommodate 収容する

mirror 反映する

chocolate browny チョコレートケーキ

ing ghrelin and increased levels of appetite-suppressing peptide YY following bariatric surgery. As recent experiments suggest, these hormones act not only on the hypothalamus but also on the reward circuit. "In the long term, we can probably mimic the effects of bariatric surgery with drugs," says Bernd Schultes of the eSwiss Medical & Surgical Center in St. Gallen, Switzerland. "That is the great dream."

In the meantime, several clinicians are using recent revelations about hedonic hunger to help people like Brien. Yi-Hao Yu, one of Brien's doctors at Greenwich Hospital in Connecticut, proposes that obesity takes at least two distinct but sometimes overlapping forms: metabolic and hedonic. Because he believes Brien struggles primarily with hedonic obesity, Yu recently prescribed the drug Victoza, which is known to reduce pleasure-driven eating. In contrast, drugs that typically target the hypothalamus would work better if a patient's underlying problem was a flaw in the body's ability to maintain a steady weight.

Drexel's Lowe, for his part, has focused on new approaches to behavior modification. "The traditional idea is that we can teach overweight people to improve their self-control," Lowe says. "The new idea is that the foods themselves are more the problem." For some people, palatable foods invoke such a strong response in the brain's reward circuit—and so dramatically alter their biology—that willpower will rarely, if ever, be sufficient to resist eating those foods once they are around. Instead, Lowe says, "we have to reengineer the food environment." In practical terms, that means never bringing fatty, super-sweet foods into the house in the first place and avoiding venues that offer them whenever possible.

Elizabeth O'Donnell has put these lessons into practice. A 53-year-old store owner who lives in Wallingford,

Pa., O'Donnell learned to modify her personal food environment at home and on the road after participating in one of Lowe's weight-loss studies. She says she is particularly helpless before sweets and pastries and so has committed to keeping them out of her home and to avoiding restaurants with all-you-can-eat dessert tables—which in the past led her to consume "an excess of 3,000 or 4,000 calories." On a recent visit to Walt Disney World, for example, she bypassed the park's many buffet-style restaurants in favor of a smaller, counter-service eatery, where she bought a salad. That's exactly the kind of simple change that can make a huge difference in the struggle to maintain a healthy weight.

Vocabulary

on the road 外出中に

pastry 焼き菓子, ケーキ類

all-you-can-eat 食べ放題の

eatery 軽食堂, 食べもの屋

ブライエン（Matthew Brien）は過去20年，食べ過ぎと格闘している。24歳のときに身長177.8cm 体重61.2kgの細身だったのに，いまの体重は104kgある。パンやパスタ，炭酸飲料，アイスクリーム，特にアーモンドやチョコレートがたっぷりの濃厚なアイスの誘惑には抗しがたい。食事量を抑える様々な減量法を試したが，どれも長続きしなかった。「ほとんど無意識に食べてしまう」という。「ディナーは終わりか，じゃデザートだ，ってな感じ。容器入りアイスをまるごと全部食べてしまう。どうにもやめられない」。

生存のためではなく快楽のために食べる行為は古くからある。だが，ある種の食べ物（特に脂肪分に富むものや甘いもの）が脳における生化学反応を変えて過食を引き起こしている仕組みが詳しくわかってきたのはここ数年だ。

こうした食欲は「快楽飢餓（ヘドニック・ハンガー）」と呼ばれるようになった。食べる必要がないのに，強い食欲が生じる。お腹はいっぱいなのに，脳はまだ飢えている状況だ。そして，先進国で肥満者の割合が急増している一因が快楽飢餓であると主張する専門家が増えている。

快楽飢餓という言葉を2007年に初めて用いたドレクセル大学の臨床心理学者ロウ（Michael Lowe）は，食欲と体重増を理解するうえで新しいアプロー

チが開けるという。「多くの過食，必要エネルギーを超えて食べている人のおそらくすべては，おいしいものを食べて過食になっている」。肥満の原因が当人の身体のカロリー燃焼能力の欠陥によるのか，それとも感情的な渇望によるのかを見極めれば，医師が最適な治療薬や行動療法を選択するのに役立つとロウはいう。

食欲解剖

以前の研究はいわゆる「代謝的空腹」に注目してきた。生理的必要に基づく空腹のことで，お腹がすいてグーグー鳴るのがこれだ。1日の途中で体内の貯蔵エネルギーを使い始めたり，いつもより体重が落ちたりすると，ホルモンの複雑なネットワークと脳の神経経路が働いて空腹感を高める。たらふく食べたり体重が増えたりすると，同じホルモン系と脳回路が食欲を抑える。

こうした代謝的空腹感をもたらしている主なホルモンと神経回路が1980年代までに解明された。空腹感を調節しているのは脳の視床下部という領域で，視床下部の神経細胞が様々なホルモンの生産を引き起こすとともに，それらのホルモンに非常に敏感に反応している。

多くの生物学的過程と同様，これらの化学シグナルは相互に連結して抑制と均衡のネットワークを作り出している。当面必要としているカロリーを超えて食べると，余剰分の一部は全身の脂肪細胞に蓄えられる。これらの脂肪細胞が成長を始めると，「レプチン」というホルモンを分泌し，これが血流に乗って脳に運ばれて，視床下部に別の一連のホルモンを放出するように伝える。それらのホルモンは食欲を抑えるとともに，余剰カロリーを燃焼するために細胞の活動を高める。こうして，すべてが元のバランスを回復する。

同様に，胃腸の細胞は食物の存在を検知すると「コレシストキニン」や「ペプチドYY」など，食欲を抑制する様々なホルモンを分泌する。それらは視床下部に移動して作用するか，あるいは迷走神経（心臓や胃腸と脳を結んでいる長い神経）に直接に働きかける。これと対照的に，胃が空っぽで血糖値が低いときに胃から放出される「グレリン」というホルモンは視床下部に逆の作用を及ぼして食欲を刺激する。

だが 1990 年代末までに，別の作用経路が脳画像研究やネズミを用いた動物実験によって明らかになってきた。快楽を求めて食べる過程の背景をなす経路だ。代謝的空腹をもたらすのと同じ多くのホルモンが関与しているようだが，代謝的空腹の場合とは違って，報酬回路というまったく異なる脳領域が活性化される。この複雑な神経回路はこれまで主に薬物依存，より最近では病的なギャンブルなどの強迫行動との関連で研究されてきたものだ。

非常に甘いものや脂肪分の豊富な食べ物は麻薬やギャンブルと同じように報酬回路をとりこにすることがわかった。こうしたカロリー豊富な食べ物は，人間が生物として進化してきた歴史の大半においては，めったにないごちそうであり，食物不足の厳しい状況下では特に強く求められた。その当時，甘いものや脂肪が手に入ったらできるだけお腹に詰め込むのが生存の手立てだった。しかし，安価で高カロリーの食物があふれた飽食の現代社会においては，この本能が裏目に出る。「十分な食物を得て餓死を避けるのが課題だったものが，現代ではまったく異なる課題に置き換わった。体重を増やさぬよう，必要以上に食べるのを避けることだ」とロウはいう。

甘く脂肪に富んだうまそうな食物を見ただけで，それを口にする前から脳の報酬回路が反応して興奮することが研究によって示された。これらの食物が舌に触れると，味蕾からの信号が脳の様々な領域に伝わり，これに反応して神経伝達物質のドーパミンが放出されて強い快感を生じる。おいしい食物の過食を頻繁に繰り返すと脳がドーパミンで飽和し，ついにはその状況に順応する「減感作」が起こる。神経細胞の表面でドーパミンを認識して反応している受容体の数が減るのだ。この結果，過食者は以前に経験したのと同じ快感を得るのに以前よりも多量の糖と脂肪が必要になる。実際，これらの人々は満足感を再び経験するために，あるいは単に維持するために，過食を続けているらしい。

視床下部に作用する食欲刺激ホルモンの一部は報酬回路にも影響を与えていることが判明した。スウェーデンにあるイエーテボリ大学の研究チームは 2007 年から 2011 年にかけての一連の研究で，胃から分泌された空腹ホルモンのグレリンが脳の報酬回路に直接作用してドーパミンの放出を増やしていることを実証した。また，グレリンがニューロンに結合するのを薬剤によって阻害すると，

肥満者の過食が減ることもわかった。

食事を続けると，通常ならレプチンとインスリン（これらは余剰カロリーを摂取すると増える）がドーパミンの放出を抑制して，快感が薄れていく。だが，体内の脂肪組織が増えると脳がこれらのホルモンに反応しなくなることが，ネズミを使った近年の研究で示唆された。こうして食べ続けることによって脳はドーパミンであふれかえり，快感を生む閾値は上がり続ける。

渇望を抑えるアプローチ

肥満者を対象に一部で実施されている一種の手術は，体重調整におけるグレリンの重要性を裏づけるとともに，多くの人が生理的な必要量をはるかに超えて食べてしまう理由について生物学的な知見を与えてくれる。「肥満外科手術」や「減量手術」として知られるこの処置はいわば肥満治療の最後の手段で，胃の一部を切除するか胃を帯できつく縛ることで容積を劇的に減らし，数十cm^3の食物しか収容できないようにする。

通常，手術から1カ月以内に空腹感が総じて弱まり，糖分や脂肪に富む食物に対する魅力が薄れる。以前よりも胃が小さくなり，胃が作り出すホルモンの量が減ったためと考えられる。この渇望の低下が神経回路の変化を映していることが，近年の脳画像研究で明らかになった。チョコレートケーキなど誘惑的な食物の画像を見たり名称を聞いたりした際の報酬回路の反応が手術後にはずっと弱まるほか，ドーパミンに対する感度が戻って少量でも反応するようになっていた。

「胃腸を解剖学的に変えることで，脳に達するホルモンの量を変えるわけだ」と，ジョンズ・ホプキンス大学医学部の外科医スティール（Kimberley Steele）はいう。肥満外科手術の後に食欲刺激ホルモンのグレリンの濃度が下がり，食欲を抑制するペプチドYYの濃度が上がることが，いくつかの研究で報告された。最近の実験が示唆しているように，これらのホルモンは視床下部だけでなく報酬回路にも作用する。「長期的には，同じ効果を薬剤によって引き出せるようになるだろう」とスイスのザンクト・ガレンにあるeスイスメディカル＆外科センターに所属するシュルテス（Bernd Schultes）はいう。「それが大きな夢だ」。

一方，快楽飢餓に関する近年の知見を役立てようとする試みも進んでいる。ブライエンの主治医であるグリニッジ病院（コネティカット州）のユ（Yi-Hao Yu）は，肥満には少なくとも2つの形態があり，それらが重なっている場合もあると提唱している。代謝的な肥満と快楽追求型の肥満だ。ユはブライエンが主に快楽追求型肥満だと考え，快感を求める摂食行動を抑えることが知られている「ビクトーザ」という薬を彼に処方した。一方，代謝機能の問題が背景になっている場合には，視床下部に作用する薬が効果的だろう。

　ドレクセル大学のロウは患者の行動を変える新アプローチに集中している。「体重過多の人に自己統制を強めるよう教えるのが伝統的な考え方だが，新しい考え方では食物そのものの問題がむしろ大きいとみる」とロウはいう。人によっては，うまそうな食物によって脳の報酬回路が非常に強く反応し，生物学的に劇的な変化が生じて，意志の力で食欲に抵抗することができなくなる。だからむしろ，「食物環境を作り替える必要がある」とロウはいう。脂肪分に富む食物や甘い食物をそもそも自宅に持ち込まないようにし，外出先でもそうした食物を提供する場を極力避けることだ。

　ペンシルベニア州ウォリンフォードに住む53歳の商店主オドネル（Elizabeth O'Donnell）はこれを実践している。彼女はロウの減量研究に被験者として参加した際に，自宅や外出先で自分の食環境を調整する方法を学んだ。甘いお菓子やケーキに特に弱いので，これらを自宅に持ち込まないと誓い，デザート食べ放題コーナーのあるレストランを避けることにした。以前は食べ放題コーナーで「3000キロカロリーか4000キロカロリーは余計に食べていた」という。最近ディズニー・ワールドに行った際にはビュッフェ形式のレストランを避け，カウンター販売の小さな店でサラダを買った。これこそ，健康な体重を保つ試みにおいて巨大な違いをもたらすための，簡単に実行できる小さな変化なのだ。

Artificial Sweeteners Get a Gut Check

人工甘味料で肥満に？

腸内細菌叢を変えている可能性がある

E. R. シェル（ジャーナリスト）

掲載：SCIENTIFIC AMERICAN April 2015, 日経サイエンス 2015 年 10 月号

Many of us, particularly those who prefer to eat our cake and look like we have not done so, have a love-hate relationship with artificial sweeteners. These seemingly magical molecules deliver a dulcet taste without its customary caloric punch. We guzzle enormous quantities of these chemicals, mostly in the form of aspartame, sucralose and saccharin, which are used to enliven the flavor of everything from Diet Coke to toothpaste. Yet there are worries. Many suspect that all this sweetness comes at some hidden cost to our health, although science has only pointed at vague links to problems.

Vocabulary
artificial sweetener 人工甘味料
dulcet 甘美な
customary 通例の
guzzle たくさん食う
aspartame アスパルテーム
sucralose スクラロース
saccharin サッカリン
enliven 活気づける

Last year, though, a team of Israeli scientists put together a stronger case. The researchers concluded from studies of mice that ingesting artificial sweeteners might lead to—of all things—obesity and related ailments such as diabetes. This study was not the first to note this link in animals, but it was the first to find evidence of a plausible cause: the sweeteners appear to change the population of intestinal bacteria that direct metabolism, the conversion of food to energy or stored fuel. And this result suggests the connection might also exist in humans.

put together まとめる, 作り上げる

ingest 摂取する
of all things こともあろうに
obesity 肥満
diabetes 糖尿病

intestinal bacteria 腸内細菌
metabolism 代謝

In humans, as well as mice, the ability to digest and extract energy from our food is determined not only by our genes but also by the activity of the trillions of microbes that dwell within our digestive tract; collectively, these bacteria are known as the gut microbiome. The Israeli study suggests that artificial sweeteners enhance the populations of gut bacteria that are more efficient at pulling energy from our food and turning that energy into fat. In other words, artificial sweeteners may favor the growth of bacteria that make more calories available to us, calories that can then find their way to our hips, thighs and midriffs, says Peter Turnbaugh of the University of California, San Francisco, an expert on the interplay of bacteria and metabolism.

Vocabulary

microbiome
▶ Technical Terms

thigh 太もも
midriff 胴の中央部

metabolism 代謝

BACTERIAL GLUTTONS

In the Israeli experiment, 10-week-old mice were fed a daily dose of aspartame, sucralose or saccharin. Another cluster of mice were given water laced with one of two natural sugars, glucose or sucrose. After 11 weeks, the mice receiving sugar were doing fine, whereas the mice fed artificial sweeteners had abnormally high blood sugar (glucose) levels, an indication that their tissues were having difficulty absorbing glucose from the blood. Left unchecked, this "glucose intolerance" can lead to a host of health problems, including diabetes and a heightened risk of liver and heart disease. But it is reversible: after the mice were treated with broad-spectrum antibiotics to kill all their gut bacteria, the microbial population eventually returned to its original makeup and balance, as did blood glucose control.

laced with ～を添加した
glucose ブドウ糖
sucrose ショ糖

absorb 吸収する
glucose intolerance ブドウ糖耐性
diabetes 糖尿病

broad-spectrum antibiotic 広域スペクトル抗生物質

Technical Terms
マイクロバイオーム(**microbiome**)　その場に生息している微生物全体のことで，この記事では腸内細菌叢(そう)と同義。オーム(-ome)は「集団」「全体」を意味する。人体には腸のほか口の中や皮膚などにも多様な微生物が常在しており，それらが人体と相互作用して様々な影響を及ぼし合っている。近年は細菌集団の DNA をまとめて解析してそこに含まれる細菌の種類などを突き止められるようになり，人によって異なるマイクロバイオームの構成や健康との関連を調べる研究が盛んになっている。

"These bacteria are not agnostic to artificial sweeteners," says computational biologist Eran Segal of the Weizmann Institute of Science in Rehovot, Israel, one of the two scientists leading the study. The investigators also found that the microbial populations that thrived on artificial sweeteners were the very same ones shown—by other researchers—to be particularly abundant in the guts of genetically obese mice.

Jeffrey Gordon, a physician and biologist at Washington University in St. Louis, has done research showing that this relation between bacteria and obesity is more than a coincidence. Gordon notes that more than 90 percent of the bacterial species in the gut come from just two subgroups—Bacteroidetes and Firmicutes. Gordon and his team found several years ago that genetically obese mice (the animals lacked the ability to make leptin, a hormone that limits appetite) had 50 percent fewer Bacteroidetes bacteria and 50 percent more Firmicutes bacteria than normal mice did. When they transferred a sample of the Firmicutes bacterial population from the obese mice into normal-weight ones, the normal mice became fatter. The reason for this response, Gordon says, was twofold: Firmicutes bacteria transplanted from the fat mice produced more of the enzymes that helped the animals extract more energy from their food, and the bacteria also manipulated the genes of the normal mice in ways that triggered the storage of fat rather than its breakdown for energy.

Gordon believes something similar occurs in obese humans. He found that the proportion of Bacteroidetes to Firmicutes bacteria increases as fat people lose weight through either a low-fat or low-carbohydrate diet. Stanford University microbiologist David Relman says this finding suggests that the bacteria in the human gut may not only influence our ability to extract calories and store energy from our diet but also have an impact on the balance of hormones, such as leptin, that shape our very eating behavior, leading some of us to eat more than others in any given situation.

The burning question, of course, is whether artificial sweeteners can truly make humans sick and fat. Segal thinks they probably do, at least in some cases. He and his team analyzed a database of 381 men and women and found that those who used artificial sweeteners were more likely than others to be overweight. They were also more likely to have impaired glucose tolerance. Obesity is, in fact, well known as a risk factor for the development of glucose intolerance as well as more severe glucose-related ailments, such as diabetes.

These patterns do not prove that the sweeteners caused the problems. Indeed, it is quite possible that overweight people are simply more likely than others to consume artificial sweeteners. But Segal's team went further, testing the association directly in a small group of lean and healthy human volunteers who normally eschewed artificial sweeteners. After consuming the U.S. Food and Drug Administration's maximum dose of saccharin over a period of five days, four of the seven subjects showed a reduced glucose response in addition to an abrupt change in their gut microbes. The three volunteers whose glucose tolerance did not dip showed no change in their gut microbes.

Although not everyone seems susceptible to this effect, the findings do warrant more research, the scientists say. The Israeli group concluded in its paper that artificial sweeteners "may have directly contributed to enhancing the exact epidemic that they themselves were intended to fight"—that is, the sweeteners may be making at least some of us heavier and more ill.

A cause-and-effect chain from sweeteners to microbes to obesity could explain some puzzles about obese

Vocabulary

burning 火急の, 焦眉の

overweight 体重過多

ailment （慢性的な）病気

lean 細身の
eschew 避ける, 慎む
U.S. Food and Drug Administration 米食品医薬品局
dose 摂取量, 投与量

susceptible 感じやすい, 影響を受けやすい
warrant 〜に値する

epidemic 流行, 蔓延

people, says New York University gastroenterologist Ilseung Cho, who researches the role of gut bacteria in human disorders. He points out that in studies, most people who switch from sugar to low-calorie sweeteners in an effort to lose weight fail to do so at the expected rate. "We've suspected for years that changes in gut bacteria may play some role in obesity," he says, although it has been hard to pinpoint this effect. But Cho adds that it is clear that "whatever your normal diet is can have a huge impact on the bacterial population of your gut, an impact that is hard to overestimate. We know that we don't see the weight-loss benefit one would expect from these nonnutritive sweeteners, and a shift in the balance of gut bacteria may well be the reason, especially a shift that results in a change in hormonal balances. A hormone is like a force multiplier—and if a change in our gut microbes has an impact on hormones that control eating, well, that would explain a lot."

MICROBES VS. GENES

Naturally there are many questions left to answer. Cathryn Nagler, a pathologist at the University of Chicago and an expert on gut bacteria and food allergies, says that the enormous genetic variations in humans make extrapolations from mice suspect. "Still, I found the data very compelling," she says of the Israeli artificial sweetener study. Relman agrees that rodent studies are not always reflective of what happens in humans. "Animal studies can point to a general phenomenon, but animals in these studies tend to be genetically identical, while in humans, lifestyle histories and genetic differences can play a very powerful role," he says. The constellation of microbes in a human body is a reflection of that body's particular history—both genetic and environmental.

"The microbiome is a component intertwined in a complex puzzle," Relman continues. "And sometimes the genetics is so strong that it will override and drive back the microbiota." Genetic variations might explain why only four of the seven saccharin-fed humans had a change in their gut bacteria, for instance, although genetics is only one of a number of possible factors. And if someone is genetically predisposed to obesity and consumes a diet that promotes that obesity, the microbes might change to take advantage of that diet, thereby amplifying the effect.

The Israeli researchers agree that it is far too soon to conclude that artificial sweeteners cause metabolic disorders, but they and other scientists are convinced that at least one—saccharin—has a significant effect on the balance of microbes in the human gut. "The evidence is very compelling," Turnbaugh says. "Something is definitely going on." Segal, for one, is taking no chances: he says that he has switched from using artificial to natural sweetener in his morning coffee.

Vocabulary

drive back 退ける
microbiota 微生物叢

predispose かかりやすくする

metabolic disorder 代謝疾患

take chances 危険を冒す

多くの人が，特にケーキを食べても太りたくないと思っている人は，人工甘味料に愛憎を抱いている。カロリーなしに甘みをもたらすこれら魔法の分子を私たちは大量に摂取している。多くはアスパルテームやスクラロース，サッカリンで，ダイエットコークから練り歯磨きまであらゆるものの味つけに使われている。だが多くの人が，人工甘味料に健康上の隠れた悪影響があるのではないかと疑ってきた。ただし科学研究で指摘された関連性はごく曖昧なものにすぎない。

ところが昨年（2016年），イスラエルの研究チームが，より明確な事例を報告した。マウスの実験に基づいて，人工甘味料の摂取がこともあろうに肥満とそれに関連する糖尿病などの疾患につながる恐れがあると結論づけている。この関連を動物実験で指摘した研究は以前にもあったのだが，今回は初めて，その原因と考えられる証拠を見つけた。人工甘味料は代謝（食べたものをエネルギーや脂肪に変える過程）を制御している腸内細菌の構成を変えるらしい。マウスだけでなくヒトでも同様と考えられるという。

マウスもヒトも食物を消化してエネルギーを引き出す能力は自分の遺伝子だけで決まっているのではなく，消化管にすみついた何兆個もの微生物の活動によっている。腸内マイクロバイオーム（腸内細菌叢）と呼ばれる細菌集団だ。イスラエルの研究は，食物からエネルギーを効率よく引き出して脂肪に変える腸内細菌の数が人工甘味料によって増えることを示唆している。腸内細菌と代謝の関係に詳しいカリフォルニア大学サンフランシスコ校のターンバウ（Peter Turnbaugh）は，食物からより多くのカロリーを引き出してお尻や太もも，胴回りに脂肪として蓄積するのを助けている細菌の成長を，人工甘味料が促進しているらしいという。

大食いバクテリア

イスラエルの研究チームは10週齢のマウスにアスパルテームかスクラロース，サッカリンを毎日与え，対照群のマウスには天然のブドウ糖またはショ糖を添加した水を与えた。11週間後，対照群は元気だったが，人工甘味料を摂取したマウスは血糖値（血液中のブドウ糖濃度）が異常に高くなっていた。身体組織が血液からブドウ糖を吸収するのが困難になっていることを示している。この「ブドウ糖耐性」を放置すると，糖尿病や肝臓疾患，心臓病のリスクが高まるなど，数々

の問題につながる。ただし回復は可能だ。マウスに広域スペクトル抗生物質を投与して腸内細菌を一掃すると，その後再び腸内マイクロバイオームが発達して最終的には以前のバランスに戻り，血糖の制御も元通り正常になった。

「腸内細菌は人工甘味料に無頓着ではない」と，研究を主導したワイツマン科学研究所の計算生物学者シーガル（Eran Segal）はいう。同研究チームはまた，人工甘味料を摂取したマウスの腸内細菌叢が，別の研究グループが遺伝的な肥満マウスの腸内に確認していた構成と非常によく似ていることも発見した。

ワシントン大学（セントルイス）の生物学者で内科医のゴードン（Jeffrey Gordon）が行った研究は，腸内細菌と肥満のこの関係が偶然ではないことを示している。腸内細菌の種の90％以上は「バクテロイデス門」と「フィルミクテス門」というたった2つのサブグループに属していると彼は指摘する。ゴードンらは数年前，遺伝子操作で肥満にしたマウス（食欲を抑えるレプチンというホルモンを作れなくしたマウス）の腸内細菌叢はバクテロイデス門の細菌が通常のマウスに比べて50％少なく，フィルミクテス門の細菌が50％多いことを発見した。これら肥満マウスのフィルミクテス門細菌のサンプルを正常な体重のマウスの腸内に移植したところ，マウスは太った。この反応には2つの原因があるとゴードンはいう。肥満マウスから移植されたフィルミクテス門の菌によって，食物から多くのエネルギーを引き出すのに寄与する一連の酵素が生産されたことと，これらの細菌が正常マウスの遺伝子発現を変化させて，体脂肪の分解・利用ではなく蓄積を引き起こしたことだ。

太った人でも同様のことが起きているとゴードンは考えている。肥満者が低脂肪食または低炭水化物食によって減量すると，腸内でフィルミクテス門細菌に対するバクテロイデス門細菌の比が高まることを彼は発見した。スタンフォード大学の微生物学者レルマン（David Relman）はこの研究から，腸内細菌は私たちが食物からカロリーを引き出してエネルギーを蓄える能力に影響しているだけでなく，私たちの摂食行動そのものを左右するレプチンなどのホルモンのバランスに影響を与え，この結果として一部の人は常に過食になっていると考えられるという。

問題の焦点は，人工甘味料が本当に人間の健康を害して太らせているのかどうかだ。シーガルは，少なくとも一部の例については，その可能性があると考えている。彼のチームはデータベースに蓄積された男女381人の健康情報を解析し，人工甘味料を使っている人は体重過多になる傾向が強いことを見いだした。望ましくないブドウ糖耐性が生じる傾向も強い。肥満はブドウ糖耐性を生むリスク要因としてよく知られており，糖尿病など血糖値に関連するさらに深刻な疾患にもつながる。

ただ，これらのパターンは人工甘味料が原因だと証明はしていない。実際，体重過多の人は単に人工甘味料の摂取量が多いだけという可能性も十分にありうる。これに対しシーガルのチームは一歩踏み込み，いつもは人工甘味料を使っていない細身の健康な人を被験者として，人工甘味料と肥満の関連性を直接に検証した。7人の被験者に米食品医薬品局（FDA）による最大摂取許容量のサッカリンを5日間与えたところ，4人はブドウ糖に対する反応が鈍ったほか，腸内細菌叢が急変した。ブドウ糖耐性が生じなかった残りの3人は，腸内細菌に変化がなかった。

全員が人工甘味料の影響を受けるわけではないらしいものの，さらに調べるべきだろうという。研究チームは論文で，人工甘味料は「それが防止するはずだった肥満の蔓延を悪化させるのに寄与しているようだ」と結論づけた。つまり，人工甘味料は少なくとも一部の人を太らせ，不健康にしている可能性がある。

人工甘味料から腸内細菌，肥満に至る因果の鎖を想定すると肥満者に関するいくつかの謎に説明がつくだろうと，腸内細菌と病気への影響を研究しているニューヨーク大学の胃腸病学者チョウ（Ilseung Cho）はいう。減量のため砂糖から低カロリー甘味料に切り替えても，ほとんどの人は期待したような減量ができずに終わっていると彼は指摘する。「腸内細菌の変化が肥満に寄与しているのではないかと，私たちはかねて疑ってきた」。この効果を突き止めるのは困難ではあるが，「日常の食事が腸内細菌の構成にいかに大きく影響しうるか，いくら強調してもしすぎにはならない」という。「ノーカロリー甘味料に期待した減量効果が見られないのは確かであり，その理由は腸内細菌のバランスの変化，特にホルモンバランスを変えるような変化が原因なのかもしれない。ホルモンは大きな戦

力だ。腸内細菌の変化が摂食を調節しているホルモンに影響していると考えれば，おおかた説明がつく」。

腸内細菌 vs. 遺伝子

当然ながら，未解決の疑問が多く残っている。シカゴ大学の病理学者で腸内細菌と食物アレルギーを専門にしているナグラー（Cathryn Nagler）は，人間は遺伝的多様性が非常に大きいのでマウスの結果をそのまま外挿するのは危ういと指摘しつつ，「それでも，このデータは非常に説得力があると思う」とイスラエルの研究を評価する。レルマンも，ネズミの実験が人間で起こることを反映しているとは限らないという。「動物実験は一般的現象を探ることができるが，今回の研究が用いた動物は遺伝的に均質だ。これに対し人間では，それまでの生活様式や遺伝的差異が非常に大きな影響力を持つ」。人体の微生物叢は，その特定の人体の歴史を反映しており，遺伝的歴史と環境的歴史の双方を含んでいる。

「マイクロバイオームは複雑なジグソーパズルだ」とレルマンは続ける。「あるときには遺伝的要因が強く効いて，微生物叢の影響を圧倒するだろう」。例えばサッカリン摂取の実験で腸内細菌叢が変化したのが 7 人のうち 4 人だけだったのは遺伝的差異で説明できるかもしれず，そうした要因はほかにもいろいろ考えうる。そして，遺伝的に肥満しやすい人が肥満を促進する食事をしている場合，腸内細菌叢がその食事に合わせて変化し，肥満につながる効果をさらに強めているのかもしれない。

イスラエルの研究チームも人工甘味料が代謝疾患の原因だと結論づけるのは早すぎると考えているが，他の科学者も含め，少なくともサッカリンは腸内細菌叢のバランスにかなりの影響を及ぼすと確信している。「非常に説得力のある証拠が出ている」とターンバウはいう。「何かが起きているのは確実だ」。シーガルは念のため，毎朝のコーヒーに入れていた人工甘味料を砂糖に切り替えた。

Can Brown Fat Defeat Obesity?
褐色脂肪で肥満を防げるか？

暖房を抑えると太らずにすむ理由

M. W. モイヤー（サイエンスライター）

掲載：SCIENTIFIC AMERICAN August 2014, 日経サイエンス 2015 年 1 月号

For most people, "fat," particularly the kind that bulges under the skin, is a four-letter word. It makes our thighs jiggle; it lingers despite our torturous attempts to eliminate it. Too much of it increases our risk for heart disease and type 2 diabetes (the most common form of the condition). For decades researchers have looked for ways to reduce our collective stores of fat because they seemed to do more harm than good.

But biology is rarely that simple. In the late 2000s several research groups independently discovered something that shattered the consensus about the absolute dangers of body fat. Scientists had long known that humans produce at least two types of fat tissue—white and brown. Each white fat cell stores energy in the form of a single large, oily droplet but is otherwise relatively inert. In contrast, brown fat cells contain many smaller droplets, as well as chestnut-colored molecular machines known as mitochondria. These organelles in turn burn up the droplets to generate heat. Babies, who have not yet developed the ability to shiver to maintain their body temperature, rely on thermogenic deposits of brown fat in the neck and around the shoulders to stay warm. Yet investigators assumed that all brown fat disappears during childhood. The new findings revealed otherwise. Adults have brown fat, too.

Vocabulary

bulge 膨らむ, 満ちている
four-letter word 四文字語（4 文字からなる卑猥な単語, 不快語）
thigh 太もも
jiggle 揺れる
type 2 diabetes 2 型糖尿病

shatter 打ち砕く
body fat 体脂肪

white fat cell 白色脂肪細胞

brown fat cell 褐色脂肪細胞

mitochondria ミトコンドリア
organelle 細胞小器官

shiver 身震いする
thermogenic 熱を発する

1　肥満＆生活習慣

Suddenly, people started throwing around terms like holy grail to describe the promise of brown fat to combat obesity. The idea was appealingly simple: if researchers could figure out how to incite the body to produce extra brown fat or somehow rev up existing brown fat, a larger number of calories would be converted into heat, reducing deposits of white fat in the process.

Brown fat proved difficult to study, however, in part because it was so hard to find in adults. In addition, some experts doubted that enough brown fat could remain in the grown-up body to make much of a difference for the obese. Finally, the easiest way to get brown fat warmed up and going is to expose people to low temperatures, which somewhat diminishes brown fat's appeal as a weight-loss tool. The more researchers learned about brown fat, the more complications and questions arose.

Now, however, the understanding of brown fat is turning a corner. Scientists have learned new ways to pinpoint its location underneath the skin. The latest evidence suggests that it can indeed reduce excess stores of fat even in the obese. Researchers have also identified compounds that can activate brown fat without the need for unpleasantly chilling temperatures. As bizarre as it sounds, fat may become an important ally in the fight against obesity.

BIG FAT COMPLEXITIES

In 2009 three different groups independently published papers in the *New England Journal of Medicine* confirming their discovery of active brown fat cells in healthy adults. Investigators spent the next five years figuring out how to study brown fat more easily and in greater detail.

The most popular method of mapping where brown fat is located under the skin has been to scan the body using combined positron-emission tomography and

Vocabulary

throw around 派手に振り回す
holy grail 探し求めてやまないもの, 聖杯
obesity 肥満
incite 刺激する
rev up 活発にする, 強める

obese 肥満の(人), 肥満者
warm up 作動できる状態になる

compound 化合物, 合成物

ally 援軍

positron-emission tomography 陽電子放出断層撮影

computed tomography (PET-CT). This technique produces highly detailed images of the body's interior but requires a costly and invasive procedure, says Paul Lee, a research officer at the Garvan Institute of Medical Research in Sydney. To perform such scans, doctors first inject patients with solutions of radioactive but benign sugar molecules. Once the mitochondria inside brown fat cells start working, they consume the radioactive sugar, which emits gamma rays that the PET part of the scan can detect. The CT scan outlines the various different types of tissue, and the combination of the two technologies identifies brown fat cells that happen to be active while overlooking all the dormant deposits.

Lee says several new methodologies are on the horizon that could make investigating brown fat in people easier and far more accurate. Scientists have, for example, devised ways of measuring brown fat with magnetic resonance imaging (MRI), a technology that uses giant magnets to harmlessly align water molecules in the body in such a way that detailed images of its different tissues are created and that is much less invasive than PET-CT scans because no injections are necessary. Another relatively inexpensive and noninvasive option is thermal imaging, which identifies hotspots of brown fat under the skin by monitoring the temperature of the overlying skin.

As tools for studying brown fat have improved, experimenters have challenged previous pessimism about its ability to help people lose weight. In a 2012 study, six men remained inactive for three hours while wearing a cold suit that circulated water with a temperature of 64.4 degrees Fahrenheit over their skin—cold enough to lower their body temperature without causing too much shivering. That way the researchers could be sure that most of the extra calories burned during those three hours were expended by brown fat cells rather than quivering muscles.

褐色脂肪で肥満を防げるか？

Vocabulary

computed tomography コンピューター断層撮影
invasive 侵襲的な

radioactive 放射性の
benign 害のない

gamma ray ガンマ線

dormant 活動休止中の

methodology 方法

magnetic resonance imaging 磁気共鳴画像法

thermal imaging 熱画像法, サーモグラフィー

challenge 正当性を疑う, 吟味する

expend 消費する
quiver 震えさせる

The volunteers burned an extra 250 calories compared with what they would have used up during three hours of inactivity at more typical indoor temperatures. Although that may not sound like a lot, an extra 250 calories a day for two weeks would consume enough energy to allow a dieter to lose a pound of fat. "Even very modest increases in metabolism over a long period can lead to significant weight reduction," says Barbara Cannon, a physiologist at the Wenner-Gren Institute for Experimental Biology in Stockholm, who was not involved in the study.

Recent experiments have also revealed that brown fat's benefits go far beyond burning calories. A 2011 study using mice found that brown fat can fuel itself with triglycerides taken from the bloodstream—exactly the kind of fatty molecules known to increase the chances of developing metabolic syndrome, a cluster of conditions that raises the risk for heart disease, stroke and diabetes. Brown fat cells also draw sugar molecules from the blood, which could help lower the risk for type 2 diabetes; chronically high levels of blood glucose wreak havoc on the body's ability to manage those levels in the first place, which in turn sets the stage for diabetes.

BEIGE POWER

Given these findings, an increasing number of scientists and biotech companies are trying to develop ways to multiply the number of brown fat cells in the body or somehow boost their activity. In addition, they are exploring the possibility of transforming white fat cells into tissue that behaves a lot like brown fat—what they call "beige" or "brite" (brown in white) fat.

Figuring out whether cool temperatures trigger the production of beige fat, in addition to revving up brown fat, seemed like a good starting point. Last year Japanese researchers asked 12 young men with lower than

Vocabulary

metabolism 代謝

triglyceride トリグリセリド

metabolic syndrome メタボリックシンドローム
stroke 脳卒中

wreak havoc on 〜に大混乱を引き起こす
sets the stage for 〜のお膳立てをする

beige fat ベージュ脂肪

average amounts of active brown fat to sit in a 63 degree F room for two hours a day for six weeks. At first, the study participants burned an average of 108 extra calories in the cold compared with more normal indoor temperatures. After six weeks, however, their bodies were burning an extra 289 calories in the cold, and PET-CT scans indicated that their beige fat activity had indeed increased. A group of similarly aged and healthy men who were not repeatedly exposed to the cold showed no change in their metabolism. The researchers think that over the six weeks low temperatures increased the activity of a gene named *UCP1*, which seems to guide the conversion of white fat into beige fat.

Don't fancy low temperatures? Investigators have identified several molecules that may be able to stimulate such "browning" of white fat without the need for cold. Two 2012 studies showed that a hormone called irisin, which is released from muscle cells after exercise, coaxes white fat to behave like brown fat. In one of these studies, researchers injected mice with a gene that tripled the levels of the hormone in the blood of mice that were obese and had dangerously high amounts of sugar in their bloodstream. The mice lost weight and regained control of their glucose levels in just 10 days.

Exercise has also been shown to increase *UCP1* activity in brown fat, making it more active. Other naturally derived browning stimulators currently under investigation include brain-derived neurotrophic factor—a molecule that usually promotes growth of neurons—and SIRT1, a protein whose purpose remains mysterious but that may help the body manage stress.

Whereas converting existing white fat into beige fat is a promising approach, some researchers, including Cannon, think it may prove more helpful to increase amounts of brown fat itself. In 2013 she and her colleagues reported that brown fat can burn at least five times more stored energy than beige fat. Ideally, Cannon says, scientists will learn how to keep stores of brown fat as large and active throughout adulthood as they are in infancy: "The goal should be to maintain brown fat forever rather than having to re-create it."

Many researchers are confident that they will eventually hit on specific brown fat–based treatments, although most admit that such interventions most likely are 10 years away at least. In the meantime, though, self-motivated individuals can start applying some of the insights about brown fat to their own lives. "There is no doubt that an unhealthy diet and sedentary lifestyle are the two chief drivers of the obesity epidemic," Lee says, but "lack of exposure to temperature variation could be a subtle contributor."

In other words, central heating has its drawbacks, in part because it may dampen brown fat's activity. No one is quite ready to suggest turning down the thermostat in winter as a way of losing weight, however—although it undoubtedly saves money on your heating bill. Whether it might also help keep you trim and ward off chronic diseases remains to be seen.

Vocabulary

hit on 思いつく, 行き当たる

sedentary 座りがちの, 運動不足の
epidemic 流行, 蔓延

drawback 欠点, 問題点

ward off 防ぐ

ほとんどの人にとって，皮膚の下にたまる脂肪は忌まわしい敵だ。脂肪のせいでデブデブになり，拷問のようなダイエットをしてもしつこく消えない。脂肪が多すぎると心臓病と 2 型糖尿病（最も一般的な糖尿病）のリスクが高まる。蓄積した脂肪は益よりも害のほうが大きいようなので，それを減らす方法が何十年も前から模索されてきた。

だが生物学がそう単純であることはまずない。2000 年代後半，体脂肪が絶対的危険であるというそれまでの一致した見方を打ち砕く発見が複数の研究グループから独立に報告された。人間の体脂肪には白色脂肪と褐色脂肪という少なくとも 2 つのタイプがあることが以前から知られている。白色脂肪細胞はエネルギーを単一の大きな油滴の形で蓄え，どちらかというと変化しにくい。これに対し褐色脂肪細胞は多数の小さな油滴のほか，栗色をしたミトコンドリアという分子機械を含み，この細胞小器官が油滴を燃やして熱を生む。　赤ちゃんは体温維持のために身体を震わせることがまだできないため，首と肩の周りに蓄積した褐色脂肪を用いて熱を発生している。その褐色脂肪は子供時代にすべて消えてなくなると考えられていた。ところが近年の新発見によると違う。成人にも褐色脂肪が存在する。

肥満抑制の決め手として褐色脂肪が急にもてはやされるようになった。考え方は実に明快だ。新たな褐色脂肪の生産を促す方法や既存の褐色脂肪の活動を刺激する方法が見つかれば，多くのカロリーが熱に変換され，その過程で白色脂肪の蓄積が減るだろう。

だが褐色脂肪の研究は難しいことが判明した。成人の褐色脂肪を特定するのがそもそも難しいのが一因だ。加えて，肥満を目に見えて変えられるほどの褐色脂肪が成人に残っているのか疑問視する専門家もいた。さらに，褐色脂肪細胞を活性化する最も簡単な方法は人体を低温にさらすことなのだが，この方法は減量の手段としての魅力をいくぶん損なってしまう。褐色脂肪について知れば知るほど，いろいろ複雑な問題が浮上した。

しかし現在，褐色脂肪に関する理解は新段階を迎えている。皮下の褐色脂肪の位置を特定する新たな方法が開発された。最新の研究によると，肥満者

でも過剰な蓄積脂肪が褐色脂肪によって減ると考えられる。また，不快な低温に身をさらさなくても，褐色脂肪を活性化できる化合物が見つかった。奇妙に聞こえるだろうが，脂肪は肥満と闘ううえでの強力な援軍になるかもしれない。

体脂肪をめぐる複雑な状況

2009年，健康な成人で褐色脂肪細胞が働いていることを3つのチームが独立に発見し，*New England Journal of Medicine*誌に発表した。その後5年間，褐色脂肪をより容易に詳しく調べる方法が研究されてきた。

体内の褐色脂肪の位置を調べるには，陽電子放出断層撮影装置（PET）と通常のコンピューター断層撮影装置（CT）を組み合わせた「PET-CT」が最も広く使われてきた。オーストラリアのシドニーにあるガーバン医学研究所の研究官リー（Paul Lee）は，この方法は人体内部の詳細な画像が得られるが，費用のかさむ侵襲的な処置が必要になるという。撮影前にまず，無害だが放射性の糖分子を溶かした溶液を注射する必要がある。脂肪細胞のミトコンドリアが働いてこの放射性の糖を消費するとガンマ線が発生，これを装置のPET部分が検出する。一方のCT部は人体組織をタイプ別に描き出し，両者を組み合わせることで，脂肪組織全体を観察しながら，そのなかで活動している褐色脂肪を特定できる。

人体の褐色脂肪をはるかに詳しく，そしてより簡単に調べられる新しい方法がいくつか見えてきたとリーはいう。例えば磁気共鳴画像装置（MRI）を使う方法だ。MRIは強力な磁石を用いて体内の水分子を整列させて様々な組織の詳細な画像を得る装置で，PET-CTと違って放射性物質を注射する必要がない。サーモグラフィーも安価で非侵襲的な代替法だ。皮膚表面の温度をモニターすることで，その下にある褐色脂肪の"ホットスポット"を特定する。

研究ツールの進歩に伴い，褐色脂肪を減量に役立てることに対する以前の悲観論が変わってきた。2012年のある研究では，6人の男性に18℃の水が循環するコールドスーツを着せ，3時間じっとしていてもらった。18℃ならガタガタ震えることなく，体温を下げられる。つまり，この3時間の間に通常よりも余計に燃焼されたカロリーは，ほとんどが筋肉の震えではなく褐色脂肪細胞によって消費されたといえる。

被験者たちがこの3時間に燃焼したカロリーは，通常の室温で安静にしていたときと比べて250キロカロリー増えた。たいした量には聞こえないかもしれないが，1日250キロカロリーを余計に消費するのを2週間続けることは，体重を1ポンド（約450g）減らすダイエットに匹敵する。「代謝を少し高めるだけでも，長期間続けるとかなりの減量に結びつく可能性がある」と，ストックホルムにあるウェンナー・グレン実験生物学研究所の生理学者キャノン（Barbara Cannon）はいう（キャノン自身はこの研究には関与していない）。

近年の実験はまた，褐色脂肪の利点がカロリー燃焼にとどまらないことを示した。マウスを用いた2011年のある研究は，褐色脂肪が血流から取り込んだトリグリセリドを燃料にできることを突き止めた。トリグリセリドはまさにメタボリックシンドローム（心臓病と脳卒中，糖尿病につながる）の危険を高めるタイプの脂肪分子だ。また褐色脂肪は血液中の糖も取り込むので，2型糖尿病のリスクを下げるのにも役立つと考えられる。血中のブドウ糖濃度が慢性的に高いと人体が血糖値を制御する能力が損なわれ，糖尿病を招く。

ベージュの力

これらの発見を受け，体内の褐色脂肪細胞を増やす方法やその活動を高める方法の開発に取り組む研究者やバイオ企業が増えている。さらに，白色脂肪細胞を褐色脂肪と似た振る舞いをする組織に変える可能性が追求されている。「ベージュ脂肪」あるいは「ブライト脂肪」（brite＝ブラウン・イン・ホワイトを意味する造語）と呼ばれるものだ。

低温が褐色脂肪の活性化に加えベージュ脂肪形成を引き起こすかどうかを調べるのが，順当な出発点だと考えられた。ある日本の研究チームは2013年，褐色脂肪の量が平均以下の若い男性12人に，温度17℃の部屋で1日2時間ずつ座って過ごす生活を6週間続けてもらった。当初，肌寒い部屋で被験者たちが燃焼するカロリーは通常の温度の部屋にいる場合に比べて平均で108キロカロリー多かった。だが6週間後にはこれが289キロカロリーに増え，PET-CTの画像からベージュ脂肪の活動が実際に高まっていることが示された。同年齢の健康な男性で寒さに繰り返しさらされなかった人は，代謝に何の変化も見られなかった。6週間の間に，*UCP1*という遺伝子の活動が低温によって高まったと研究チームは考

えている。白色脂肪のベージュ脂肪への転換を指揮していると思われる遺伝子だ。

寒いのはいや？　大丈夫，寒さなしに白色脂肪の"褐色化"を刺激できそうな分子がいくつか特定されている。2012年の2つの研究は，運動した後に筋肉細胞から放出されるイリシン（アイリシン）というホルモンが，白色脂肪を褐色脂肪のように振る舞わせることを示した。片方の研究では，肥満して血糖値が危険なまでに上がっているマウスに，イリシンの血中濃度を3倍に増やす遺伝子を注射した。このマウスはたった10日で体重が減り，血糖値の制御を取り戻した。

運動も褐色脂肪中のUCP1遺伝子の活性を高め，褐色脂肪の働きを活発化することが示されている。研究中の自然の活性化因子にはほかに，脳由来神経栄養因子（BDNF，ニューロンの成長を促進している分子）やSIRT1というタンパク質がある。SIRT1の使命はまだ謎だが，人体がストレスに対処するのを助けているらしい。

既存の白色脂肪をベージュ脂肪に転換するのは有望なアプローチではあるが，キャノンを含め一部の研究者は褐色脂肪自体の量を増やすほうが有効だろうと考えている。2013年，キャノンらは褐色脂肪がベージュ脂肪に比べ5倍以上の蓄積エネルギーを燃焼しうると報告した。大人になっても子供時代と同様の量と活性の褐色脂肪を保てるようにする方法を見つけるのが理想だとキャノンはいう。「褐色脂肪を再び作るのではなく，ずっと持ち続けるようにするのが望ましい」。

多くの研究者は褐色脂肪に基づく肥満治療法がいずれは発案されると確信しているが，少なくとも10年先になるとみる。だが，褐色脂肪に関する知見の一部を自分の生活に生かす試みは始めようと思えばすぐにも可能だ。「不健康な食事と運動不足が肥満の2大原因であることに疑いはないが，温度変化にさらされる経験が欠如していることも肥満に多少は寄与しているのだろう」とリーはいう。

言い換えれば，セントラルヒーティングは褐色脂肪の活動を邪魔している可能性がある。しかし，減量の方法として冬場の室内設定温度を下げようという人はまずいない。暖房費の節約にはなるけれど，それが肥満と慢性疾患の防止にも役立つかどうかは，まだはっきりしていない。

Gut Reactions
腸内細菌と肥満

細身になるか太るかを腸内細菌叢が左右しているらしい

C. ウォリス（フリーランスライター）

掲載：SCIENTIFIC AMERICAN June 2014, 日経サイエンス 2014 年 10 月号

For the 35 percent of American adults who do daily battle with obesity, the main causes of their condition are all too familiar: an unhealthy diet, a sedentary lifestyle and perhaps some unlucky genes. In recent years, however, researchers have become increasingly convinced that important hidden players literally lurk in human bowels: billions on billions of gut microbes.

Throughout our evolutionary history, the microscopic denizens of our intestines have helped us break down tough plant fibers in exchange for the privilege of living in such a nutritious broth. Yet their roles appear to extend beyond digestion. New evidence indicates that gut bacteria alter the way we store fat, how we balance levels of glucose in the blood, and how we respond to hormones that make us feel hungry or full. The wrong mix of microbes, it seems, can help set the stage for obesity and diabetes from the moment of birth.

Fortunately, researchers are beginning to understand the differences between the wrong mix and a healthy one, as well as the specific factors that shape those differences. They hope to learn how to cultivate this inner ecosystem in ways that could prevent—and possibly treat—obesity, which doctors define as having a particular ratio of height and weight, known as the body mass index,

Vocabulary

obesity 肥満
sedentary 座りがちの, 運動不足の

lurk 潜む
gut microbe 腸内細菌, 腸内微生物

denizen 居住者, 生息者
intestine 腸
privilege 特権
broth 肉汁, 培養液

digestion 消化
glucose ブドウ糖, グルコース

diabetes 糖尿病

ecosystem 生態系

body mass index 体格指数

1　肥満&生活習慣

that is greater than 30. Imagine, for example, foods, baby formulas or supplements devised to promote virtuous microbes while suppressing the harmful types. "We need to think about designing foods from the inside out," suggests Jeffrey Gordon of Washington University in St. Louis. Keeping our gut microbes happy could be the elusive secret to weight control.

AN INNER RAIN FOREST

Researchers have long known that the human body is home to all manner of microorganisms, but only in the past decade or so have they come to realize that these microbes outnumber our own cells 10 to one. Rapid gene-sequencing techniques have revealed that the biggest and most diverse metropolises of "microbiota" reside in the large intestine and mouth, although impressive communities also flourish in the genital tract and on our skin.

Each of us begins to assemble a unique congregation of microbes the moment we pass through the birth canal, acquiring our mother's bacteria first and continuing to gather new members from the environment throughout life. By studying the genes of these various microbes—collectively referred to as the microbiome—investigators have identified many of the most common residents, although these can vary greatly from person to person and among different human populations. In recent years researchers have begun the transition from mere census taking to determining the kind of jobs these minute inhabitants fill in the human body and the effect they have on our overall health.

An early hint that gut microbes might play a role in obesity came from studies comparing intestinal bacteria in obese and lean individuals. In studies of twins who were both lean or both obese, researchers found that the gut community in lean people was like a rain forest brimming with many species but that the community in

Vocabulary

formula 調合乳
virtuous 有徳の
from the inside out 徹底的に

elusive つかまえにくい
secret 秘訣, 極意

gene-sequencing 遺伝子配列解析
microbiota 微生物叢
large intestine 大腸
genital tract 生殖管

congregation 集まり
birth canal 産道

microbiome マイクロバイオーム, 微生物叢
▶ 18 ページ **Technical Terms**

census taking 国勢調査をすること

obese 肥満の
lean やせた, 細身の

brim with 〜で満ちあふれる

obese people was less diverse—more like a nutrient-overloaded pond where relatively few species dominate. Lean individuals, for example, tended to have a wider variety of Bacteroidetes, a large tribe of microbes that specialize in breaking down bulky plant starches and fibers into shorter molecules that the body can use as a source of energy.

Documenting such differences does not mean the discrepancies are responsible for obesity, however. To demonstrate cause and effect, Gordon and his colleagues conducted an elegant series of experiments with so-called humanized mice, published last September in *Science*. First, they raised genetically identical baby rodents in a germ-free environment so that their bodies would be free of any bacteria. Then they populated their guts with intestinal microbes collected from obese women and their lean twin sisters (three pairs of fraternal female twins and one set of identical twins were used in the studies). The mice ate the same diet in equal amounts, yet the animals that received bacteria from an obese twin grew heavier and had more body fat than mice with microbes from a thin twin. As expected, the fat mice also had a less diverse community of microbes in the gut.

Gordon's team then repeated the experiment with one small twist: after giving the baby mice microbes from their respective twins, they moved the animals into a shared cage. This time both groups remained lean. Studies showed that the mice carrying microbes from the obese human had picked up some of their lean roommates' gut bacteria—especially varieties of Bacteroidetes—probably by consuming their feces, a typical, if unappealing, mouse behavior. To further prove the point, the researchers transferred 54 varieties of bacteria from some lean mice to those with the obese-type community of germs and found that the animals that had been destined to become obese developed a healthy weight instead. Transferring just 39

Vocabulary

Bacteroidetes バクテロイデス門
starch デンプン

discrepancy 不一致, 違い
cause and effect 因果関係
humanized mice ヒト化マウス

fraternal twin 二卵性双生児
identical twin 一卵性双生児

twist ひとひねり

feces 糞便
unappealing 魅力のない

strains did not do the trick. "Taken together, these experiments provide pretty compelling proof that there is a cause-and-effect relationship and that it was possible to prevent the development of obesity," Gordon says.

Gordon theorizes that the gut community in obese mice has certain "job vacancies" for microbes that perform key roles in maintaining a healthy body weight and normal metabolism. His studies, as well as those by other researchers, offer enticing clues about what those roles might be. Compared with the thin mice, for example, Gordon's fat mice had higher levels in their blood and muscles of substances known as branched-chain amino acids and acylcarnitines. Both these chemicals are typically elevated in people with obesity and type 2 diabetes.

Another job vacancy associated with obesity might be one normally filled by a stomach bacterium called *Helicobacter pylori*. Research by Martin Blaser of New York University suggests that it helps to regulate appetite by modulating levels of ghrelin—a hunger-stimulating hormone. *H. pylori* was once abundant in the American digestive tract but is now rare, thanks to more hygienic living conditions and the use of antibiotics, says Blaser, author of a new book entitled *Missing Microbes*.

Diet is an important factor in shaping the gut ecosystem. A diet of highly processed foods, for example, has been linked to a less diverse gut community in people. Gordon's team demonstrated the complex interaction among food, microbes and body weight by feeding their humanized mice a specially prepared unhealthy chow that was high in fat and low in fruits, vegetables and fiber (as opposed to the usual high-fiber, low-fat mouse kibble). Given this "Western diet," the mice with obese-type microbes proceeded to grow fat even when housed with lean cagemates. The unhealthy diet somehow prevented the virtuous bacteria from moving in and flourishing.

Vocabulary

compelling 説得力のある, 強力な

job vacancy 仕事の空き

metabolism 代謝
enticing 心をひく, 興味深い

branched-chain amino acid 分枝鎖アミノ酸
acylcarnitine アシルカルニチン

Helicobacter pylori ピロリ菌（ヘリコバクター・ピロリ）

ghrelin グレリン

hygienic 衛生的な
antibiotic 抗生物質

ecosystem 生態系

chow 食べもの, 食事

kibble 粗びきの穀物

腸内細菌と肥満

The interaction between diet and gut bacteria can predispose us to obesity from the day we are born, as can the mode by which we enter the world. Studies have shown that both formula-fed babies and infants delivered by cesarean section have a higher risk for obesity and diabetes than those who are breast-fed or delivered vaginally. Working together, Rob Knight of the University of Colorado Boulder and Maria Gloria Dominguez-Bello of N.Y.U. have found that as newborns traverse the birth canal, they swallow bacteria that will later help them digest milk. C-section babies skip this bacterial baptism. Babies raised on formula face a different disadvantage: they do not get substances in breast milk that nurture beneficial bacteria and limit colonization by harmful ones. According to a recent Canadian study, babies drinking formula have bacteria in their gut that are not seen in breast-fed babies until solid foods are introduced. Their presence before the gut and immune system are mature, says Dominguez-Bello, may be one reason these babies are more susceptible to allergies, asthma, eczema and celiac disease, as well as obesity.

A new appreciation for the impact of gut microbes on body weight has intensified concerns about the profligate use of antibiotics in children. Blaser has shown that when young mice are given low doses of antibiotics, similar to what farmers give livestock, they develop about 15 percent more body fat than mice that are not given such drugs. Antibiotics may annihilate some of the bacteria that help us maintain a healthy body weight. "Antibiotics are like a fire in the forest," Dominguez-Bello says. "The baby is forming a forest. If you have a fire in a forest that is new, you get extinction." When Laurie Cox, a graduate student in Blaser's laboratory, combined a high-fat diet with the antibiotics, the mice became obese. "There's a synergy," Blaser explains. He notes that antibiotic use varies greatly from state to state in the U.S., as does the prevalence of obesity, and intriguingly, the two maps line up—with both

Vocabulary

predispose 素因を作る, かかりやすくする

cesarean section 帝王切開

breast-fed 母乳で育てられた

baptism 洗礼

colonization 住み着くこと, 定着

solid food 固形食
immune system 免疫系
asthma 喘息
eczema 湿疹
celiac disease セリアック病

appreciation 認識, 理解

profligate 浪費の激しい

dose 投与量, 用量

annihilate 一掃する

synergy 相乗効果

prevalence 有病率

rates highest in parts of the South.

BEYOND PROBIOTICS

Many scientists who work on the microbiome think their research will inspire a new generation of tools to treat and prevent obesity. Still, researchers are quick to point out that this is a young field with far more questions than answers. "Data from human studies are a lot messier than the mouse data," observes Claire Fraser of the University of Maryland, who is studying obesity and gut microbes in the Old Order Amish population. Even in a homogeneous population such as the Amish, she says, there is vast individual variation that makes it difficult to isolate the role of microbiota in a complex disease like obesity.

Even so, a number of scientists are actively developing potential treatments. Dominguez-Bello, for example, is conducting a clinical trial in Puerto Rico in which babies born by cesarean section are immediately swabbed with a gauze cloth laced with the mother's vaginal fluids and resident microbes. She will track the weight and overall health of the infants in her study, comparing them with C-section babies who did not receive the gauze treatment.

A group in Amsterdam, meanwhile, is investigating whether transferring feces from lean to overweight people will lead to weight loss. U.S. researchers tend to view such "fecal transplants" as imprecise and risky. A more promising approach, says Robert Karp, who oversees National Institutes of Health grants related to obesity and the microbiome, is to identify the precise strains of bacteria associated with leanness, determine their roles and develop treatments accordingly. Gordon has proposed enriching foods with beneficial bacteria and any nutrients needed to establish them in the gut—a science-based version of today's probiotic yogurts. No one in the field

Vocabulary

messy 混乱した

Old Order Amish オールド・オーダー・アーミッシュ
homogeneous 均質な

clinical trial 臨床試験
swab こすりつける

overweight 体重過多
fecal transplants 糞便移植
imprecise 不適切

strain 系統, 菌株

establish 定着させる

believes that probiotics alone will win the war on obesity, but it seems that, along with exercising and eating right, we need to enlist our inner microbial army.

Vocabulary

enlist 協力を引き出す,動員する

肥満の原因はよく知られている。不健康な食事，運動不足，そしていくぶんかの遺伝的要因だ。だが近年の研究で，重要な影のプレーヤーが文字通り私たちのお腹に潜んでいると考えられるようになってきた。莫大な数の腸内細菌だ。

ヒトの進化史を通じて，これらの微生物は硬い植物性繊維の分解を助けることと引き換えに，栄養豊かな腸内にすみ着く特権を得た。だが，その役割は消化を助けることだけではないようだ。人体が脂肪を蓄積する方法や，血糖値を調節する仕方，さらには空腹感や満腹感をもたらすホルモンに対する反応を腸内細菌が変えていることを示す証拠が得られている。腸内細菌の構成が不適切だと肥満や糖尿病になりやすくなるようで，この影響は人の誕生の瞬間から生じうる。

幸い，不適切な腸内細菌叢と健康な細菌叢の違いや，そうした差をもたらす要因がわかってきた。この内なる生態系を上手に育めば肥満を防止（さらには治療）できるとみて，研究者たちはその方法を探っている。善玉菌を育て悪玉菌を抑える食物や乳幼児用の調合乳，栄養補助食品が考えられる。「食物を徹底的に考え直す必要がある」とワシントン大学（セントルイス）のゴードン（Jeffrey Gordon）はいう。腸内細菌を健全に保つことが体重管理の秘訣なのかもしれない。

お腹のなかの熱帯雨林

様々な微生物が人体にすみ着いていることは昔から知られていたが，その数が本人の細胞の10倍に達することが判明したのは10年ほど前だ。生殖管と皮膚にも驚くほどたくさんいるが，最大にして最も多様な微生物叢は大腸と口内に存在することが，高速の遺伝子解析によって明らかになった。

人間は産道を通過して生まれ落ちた瞬間から，その人に独特な微生物を集め始める。最初に獲得するのは母親が持っていた細菌で，その後は一生を通じて環境中から新たな微生物を獲得し続ける。これらマイクロバイオームと総称

される微生物叢の遺伝子を調べた結果，誰もが共通して持っている多くの微生物がわかった。とはいえ，それらは人によって，あるいは人口集団によって大きく異なる。近年の研究は単に微生物叢のメンバーを特定するだけではなく，それら小さな居住者が人体で果たしている役割と健康全般への影響を見極めることに力点が移ってきた。

腸内細菌が肥満に関係している可能性は，太った人とやせた人の腸内細菌叢の比較から示唆された。やせた双子と太った双子を調べた結果，やせた双子の腸内細菌叢は多くの生物種がいる熱帯雨林に似ているのに，太った双子の細菌叢は多様性が乏しく，少数の種しかいない富栄養化した池に近いことがわかった。例えばやせた人にはバクテロイデス門に属する多様な細菌がいる。これらの細菌は植物が含むデンプンや繊維などの大きな分子を分解して，人体がエネルギー源としている小さな分子に変える能力に優れている。

そうした違いが肥満の原因なのか，ゴードンらは因果関係を確かめるため，いわゆる「ヒト化マウス」を用いた一連の実験を行い，2013年9月に*Science*誌に報告した。まず遺伝的に同一なマウスを無菌環境で育て，体内に一切の細菌が存在しない状態にした。次いで肥満女性とその双子の姉妹であるやせた女性から採取した腸内細菌を，これら無菌マウスの腸に移植した（二卵性双生児3組と，一卵性双生児1組が細菌を提供）。マウスは同量の同じ餌を与えられたが，太った人から細菌をもらいうけたマウスはやせたほうから細菌をもらったマウスに比べ体重が重く，体脂肪量も多くなった。また予想通り，肥満マウスの腸内細菌叢は多様性が乏しかった。

ゴードンらはこの実験に少し手を加えて繰り返した。マウスにそれぞれの細菌を移植した後，同じケージで育てたのだ。すると，どちらの細菌を移植されたマウスも肥満しなかった。肥満女性の細菌を移植されたマウスが，同じケージにいるやせたマウスの腸内細菌の一部，特にバクテロイデス門の細菌を，糞を食べることによって獲得したようだ（糞を食べるのはマウスでは一般的な行動）。これを確かめるため，やせ型マウスの腸内から54種の細菌を取り出して肥満タイプの細菌叢を持つマウスに移植したところ，適切体重になった。移植する細菌を39種にとどめると，この変化は生じなかった。「因果関係の存在を示す強力な証

明だ。肥満の進行を防げるだろう」とゴードンはいう。

肥満マウスの腸内細菌叢は健康な体重維持と正常な代謝に重要な役割を果たす微生物を欠いているのだとゴードンはみる。その役割とはどんなものなのか，興味深い手がかりが得られている。例えばゴードンの肥満マウスはやせたマウスに比べ，血液中と筋肉中に「分枝鎖アミノ酸」とアシルカルニチンという物質が多かった。これらの物質はどちらも，肥満者や2型糖尿病患者で高くなることが知られている。

また肥満タイプの細菌叢は，通常なら胃にいるピロリ菌が果たしている役割が欠けているのかもしれない。ニューヨーク大学のブレーザー（Martin Blaser）はピロリ菌が空腹感を刺激するグレリンというホルモンの濃度を変えることで食欲を調節している可能性を示した。ピロリ菌はかつて米国人の消化管によく見られたが，現在では衛生的な生活環境と抗生物質使用のためにまれになっているとブレーザーはいう。

腸内生態系の形成という点で，食事は重要な要因だ。例えば加工度の高い食物は多様性の乏しい腸内細菌叢と関連づけられている。ゴードンらは特別に調合した不健康食をヒト化マウスに与え，食物と微生物，体重の間の複雑な相互作用を明らかにした。本来は繊維質の多い低脂肪の餌がふつうだが，野菜や果物，繊維が少なく脂肪の多い餌を食べさせた。肥満タイプの細菌叢を持つマウスにこうした"西欧風"の食事を与えると，やせ型マウスと同じケージで育てても，体脂肪の蓄積が進んだ。不健康な食事によって，善玉微生物の移転と繁殖が妨げられたようだ。

食事と腸内細菌のこうした相互作用は，人の誕生の瞬間から肥満しやすさを左右している可能性がある。そして出産の方法にもよる。調合乳で育った乳児と帝王切開で生まれた赤ん坊は，母乳育ちや自然分娩で生まれた子に比べ肥満や糖尿病になるリスクが高いことが示されている。コロラド大学ボールダー校のナイト（Rob Knight）とニューヨーク大学のドミンゲス=ベッロ（Maria Gloria Dominguez-Bello）は共同で，新生児が産道を通過する際に細菌をのみ込み，この細菌が後に乳の消化を助けていることを見いだした。帝王切開で生まれた子は

この細菌の洗礼を受けていない。調合乳育ちの乳児には別のハンデがある。母乳には有益な微生物を育て有害な微生物の定着を制限する物質が含まれているのだが，それを得られないのだ。カナダでの最近の研究によると，調合乳育ちの乳児は固形食を食べ始めるまで，母乳育ちの乳児にはない細菌が腸内に存在していた。腸や免疫系が未熟な段階でそうした細菌が存在することが，調合乳育ちの子が肥満やアレルギー，喘息，湿疹，セリアック病になりやすい一因なのかもしれないとドミンゲス＝ベッロはいう。

腸内細菌が肥満に及ぼす影響が認識されたことで，小児に対する抗生物質の野放図な使用に懸念が高まっている。畜産農家が家畜に与えているのと同レベルの少量の抗生物質を子マウスに与えると，体脂肪の蓄積が約15％増えることをブレーザーは示した。健康体重の維持に寄与している細菌の一部が抗生物質で一掃されたのかもしれない。「乳児は腸内細菌叢という森の形成途上にあり，抗生物質は山火事のようなもの」とドミンゲス＝ベッロはいう。「できかけの森が火事になったら，すべてが絶滅だ」。ブレーザーが指導する大学院生のコックス（Laurie Cox）が高脂肪食と抗生物質を組み合わせてマウスに与えたところ，肥満になった。「一種の相乗効果だ」とブレーザーは説明する。米国では州によって抗生物質の使用量が大きく異なり，肥満者の率も違うが，興味深いことに両者は重なると彼は指摘する。米国南部ではこれらがともに高い。

プロバイオティック補助食品を超えて

マイクロバイオームを研究する科学者の多くは，その研究が肥満を治療・予防する新手段につながると考えている。だが，この分野はまだ若く，答えよりも疑問のほうがはるかに多いともいう。「ヒトに関するデータはマウスの実験データに比べはるかに混乱している」とメリーランド大学のフレーザー（Claire Fraser）はみる。彼女はオールド・オーダー・アーミッシュという特定の人口集団を対象に肥満と腸内細菌を調べているが，こうした均質な人口集団でさえ腸内細菌叢の個人差は大きく，肥満など複雑な病気に細菌叢が果たしている役割を分離するのは難しいという。

それでも，治療法の可能性を熱心に追求している科学者はいる。例えばドミンゲス＝ベッロは，帝王切開で生まれた新生児に母親の膣の体液と微生物

を染み込ませたガーゼをすぐにこすりつけて，その影響を観察する試験をプエルトリコで実施中だ。この処置を受けていない帝王切開児と，健康状態全般を追跡して比較する。

　一方，あるオランダの研究チームは，やせた人の糞便を体重過多の人の腸内に注入すると減量につながるかどうかを調べている。ただ，米国の研究者はこうした「糞便移植」を不適切で危険であるとみる傾向が強い。肥満とマイクロバイオームに関する米国立衛生研究所（NIH）の研究助成金を監督しているカープ（Robert Karp）は，やせ型の体型に伴う細菌を特定してその役割を突き止め，治療法を開発するほうがよいという。ゴードンは有益な細菌をそれが腸内に定着するのに必要とする栄養分とともに食品に添加することを提案している。いわば科学的プロバイオティックヨーグルトだ。現在のプロバイオティック補助食品だけでは肥満との戦いに勝てないかもしれないが，運動や正しい食事とともに内なる微生物軍団の力を引き出すことが必要なようだ。

Is Fasting Good for You?
プチ断食は健康によい？

間欠的断食で健康になれるかもしれない。ただし裏づけとなるデータはまだ希薄だ

D. スティップ（サイエンスライター）

掲載：SCIENTIFIC AMERICAN January 2013, 日経サイエンス 2013 年 8 月号

In E. B White's beloved novel *Charlotte's Web*, an old sheep advises the gluttonous rat Templeton that he would live longer if he ate less. "Who wants to live forever?" Templeton sneers. "I get untold satisfaction from the pleasures of the feast."

It is easy to empathize with Templeton, but the sheep's claim has some merit. Studies have shown that reducing typical calorie consumption, usually by 30 to 40 percent, extends life span by a third or more in many animals, including nematodes, fruit flies and rodents. When it comes to calorie restriction in primates and people, however, the jury is still out. Although some studies have suggested that monkeys that eat less live longer, a new 25-year-long primate study concluded that calorie restriction does not extend average life span in rhesus monkeys. Even if calorie restriction does not help anyone live longer, a large portion of the data supports the idea that limiting food intake reduces the risks of diseases common in old age and lengthens the period of life spent in good health.

If only one could claim those benefits without being hungry all the time. There might be a way. In recent years researchers have focused on a strategy known as intermittent fasting as a promising alternative to continuous calorie restriction.

Vocabulary
gluttonous 食いしん坊の

sneer せせら笑う
feast ごちそう

empathize 共感する

consumption 摂取
nematode 線虫
fruit fly ショウジョウバエ
rodent 齧歯（げっし）類
primate 霊長類
the jury is still out 結論はまだ出ていない

rhesus monkey アカゲザル

intermittent fasting 間欠的断食

Intermittent fasting, which includes everything from periodic multiday fasts to skipping a meal or two on certain days of the week, may promote some of the same health benefits that uninterrupted calorie restriction promises. The idea of intermittent fasting is more palatable to most people because, as Templeton would be happy to hear, one does not have to renounce the pleasures of the feast. Studies indicate that rodents that feast one day and fast the next often consume fewer calories overall than they would normally and live just as long as rats eating calorie-restricted meals every single day.

In a 2003 mouse study overseen by Mark Mattson, head of the National Institute on Aging's neuroscience laboratory, mice that fasted regularly were healthier by some measures than mice subjected to continuous calorie restriction; they had lower levels of insulin and glucose in their blood, for example, which signified increased sensitivity to insulin and a reduced risk of diabetes.

THE FIRST FASTS

Religions have long maintained that fasting is good for the soul, but its bodily benefits were not widely recognized until the early 1900s, when doctors began recommending it to treat various disorders—such as diabetes, obesity and epilepsy.

Related research on calorie restriction took off in the 1930s, after Cornell University nutritionist Clive McCay discovered that rats subjected to stringent daily dieting from an early age lived longer and were less likely to develop cancer and other diseases as they aged, compared with animals that ate at will. Research on calorie restriction and periodic fasting intersected in 1945, when University of Chicago scientists reported that alternate-day feeding extended the life span of rats as much as daily dieting in McCay's earlier experiments. Moreover, intermittent fasting "seems to delay the development of the disorders that lead to death," the Chicago researchers wrote.

In the next decades research into antiaging diets took a backseat to more influential medical advances, such as the continued development of antibiotics and coronary artery bypass surgery. More recently, however, Mattson and other researchers have championed the idea that intermittent fasting probably lowers the risks of degenerative brain diseases in later life. Mattson and his colleagues have shown that periodic fasting protects neurons against various kinds of damaging stress, at least in rodents. One of his earliest studies revealed that alternate-day feeding made the rats' brains resistant to toxins that induce cellular damage akin to the kind cells endure as they age. In follow-up rodent studies, his group found that intermittent fasting protects against stroke damage, suppresses motor deficits in a mouse model of Parkinson's disease and slows cognitive decline in mice genetically engineered to mimic the symptoms of Alzheimer's. A decidedly slender man, Mattson has long skipped breakfast and lunch except on weekends. "It makes me more productive," he says. The 55-year-old researcher, who has a Ph.D. in biology but not a medical degree, has written or co-authored more than 700 articles.

Mattson thinks that intermittent fasting acts in part as a form of mild stress that continually revs up cellular defenses against molecular damage. For instance, occasional fasting increases the levels of "chaperone proteins," which prevent the incorrect assembly of other molecules in the cell. Additionally, fasting mice have higher levels of brain-derived neurotrophic factor (BDNF), a protein that prevents stressed neurons from dying. Low levels of BDNF have been linked to everything from depression to Alzheimer's, although it is still unclear whether these findings reflect cause and effect. Fasting also ramps up autophagy, a kind of garbage-disposal

Vocabulary

take a backseat to 〜に譲る, 二番手になる
antibiotic 抗生物質
coronary artery bypass surgery 冠動脈バイパス手術

degenerative brain disease 脳変性疾患

neuron ニューロン, 神経細胞

stroke 脳卒中
motor deficit 運動障害
cognitive 認知力の

decidedly 明らかに

rev up 活発にする, 強める

chaperone protein シャペロン・タンパク質

brain-derived neurotrophic factor 脳由来神経栄養因子

autophagy
▶ Technical Terms

system in cells that gets rid of damaged molecules, including ones that have been previously tied to Alzheimer's, Parkinson's and other neurological diseases.

One of intermittent fasting's main effects seems to be increasing the body's responsiveness to insulin, the hormone that regulates blood sugar. Decreased sensitivity to insulin often accompanies obesity and has been linked to diabetes and heart failure; long-lived animals and people tend to have unusually low insulin, presumably because their cells are more sensitive to the hormone and therefore need less of it. A recent study at the Salk Institute for Biological Studies in La Jolla, Calif., showed that mice that feasted on fatty foods for eight hours a day and subsequently fasted for the rest of each day did not become obese or show dangerously high insulin levels.

The idea that periodic fasting may offer some of the same health benefits as continuous calorie restriction—and allows for some feasting while slimming down—has convinced an increasing number of people to try it, says Steve Mount, a University of Maryland genetics professor who has moderated a Yahoo discussion group on intermittent fasting for more than seven years. Intermittent fasting "isn't a panacea—it's always hard to lose weight," adds Mount, who has fasted three days a week since 2004. "But the theory [that it activates the same signaling pathways in cells as calorie restriction] makes sense."

ON THIN GROUND

Despite the growing enthusiasm for intermittent fasting, researchers have conducted few robust clini-

Vocabulary

insulin インスリン

panacea 万能薬

Technical Terms

オートファジー（**autophagy**）　細胞が自分の細胞質に生じた異常なタンパク質などを小胞に集積して消化・分解すること。分解産物は捨てられずに再利用される。自分を"食べて"次の成長に役立てているわけで「自食作用」とも呼ばれる。この仕組みを解明した大隅良典博士に2016年のノーベル生理学・医学賞が贈られたのは記憶に新しい。

cal trials, and its long-term effects in people remain uncertain. Still, a 1956 Spanish study sheds some light, says Louisiana-based physician James B. Johnson, who co-authored a 2006 analysis of the study's results. In the Spanish study, 60 elderly men and women fasted and feasted on alternate days for three years. The 60 participants spent 123 days in the infirmary, and six died. Meanwhile 60 nonfasting seniors racked up 219 infirmary days, and 13 died.

In 2007 Johnson, Mattson and their colleagues published a clinical study showing a rapid, significant alleviation of asthma symptoms and various signs of inflammation in nine overweight asthmatics who near-fasted every other day for two months.

Detracting from these promising results, however, the literature on intermittent fasting also includes several red flags. A 2011 Brazilian study in rats suggests that long-term intermittent fasting increases blood glucose and tissue levels of oxidizing compounds that could damage cells. Moreover, in a 2010 study co-authored by Mattson, periodically fasting rats mysteriously developed stiff heart tissue, which in turn impeded the organ's ability to pump blood.

And some weight-loss experts are skeptical about fasting, citing its hunger pangs and the possible dangers of compensatory gorging. Indeed, the most recent primate study on calorie restriction—the one that failed to extend life span—underscores the need for caution when radically altering the way people eat.

Still, from an evolutionary perspective, three meals a day is a strange modern invention. Volatility in our ancient ancestors' food supplies most likely brought on frequent fasting—not to mention malnutrition and starvation. Yet Mattson believes that such evolutionary pressures selected

Vocabulary

clinical trial 臨床試験

infirmary 養護施設

alleviation 軽減, 緩和
asthma 喘息
inflammation 炎症
asthmatic 喘息患者

detract 価値を減ずる

oxidizing compound 酸化性物質

pang 苦しみ
gorging 大食い

volatility 不安定で, はかないこと
malnutrition 栄養不良
starvation 飢餓

for genes that strengthened brain areas involved in learning and memory, which increased the odds of finding food and surviving. If he is right, intermittent fasting may be both a smart and smartening way to live.

Vocabulary

ホ　ワイト作の『シャーロットのおくりもの』で，食いしん坊ネズミのテンプルトンに羊が「食べる量を減らしたら長生きできるよ」と助言する。テンプルトンは「いったい誰が永遠に生きたいなんて思うのかね？」とせせら笑う。「ごちそうのおかげでどれだけ幸せか，言葉では言い尽くせない」。

そ　れはもっともではあるが，羊の助言にも一理ある。カロリー摂取を 30 〜 40％減らすと寿命が 3 割以上延びることが，線虫やショウジョウバエ，齧歯（げっし）類など多くの動物で示されている。だが人間を含め霊長類で同じ効果が生じるかどうかは，はっきりしていない。一部の研究は食事を減らしたサルが長生きすることを示唆しているものの，アカゲザルについてカロリー制限の影響を 25 年にわたって調べた最近の研究は，平均寿命は延びないと結論づけた。ただ，長生きにはつながらなくても，カロリー制限が高齢者によく見られる病気の発症を抑え，健康に過ごせる期間を延ばすことを裏づけるデータは数多く蓄積している。

空　腹感なしにそうした恩恵を得られればよいのだが……。それが可能かもしれない。「間欠的断食」という方法が注目されている。

間　欠的断食（訳注：日本では俗に「プチ断食」とも呼ばれている）は 2 日以上の断食を周期的に繰り返すものから，週の決まった曜日に 1 食または 2 食を抜くものまで様々だが，連続的カロリー制限と同じ健康上の利点が得られる可能性がある。おいしい食事を放棄しなくてもよいので，多くの人にとってカロリー制限よりも好ましい。ラットを使った実験で，食事と絶食を 1 日交替で繰り返すとカロリー摂取量が総じて少なくなるほか，カロリー制限を毎日続けたラットと同様の長生きになることが示された。

国立加齢研究所（NIA）の神経科学研究室長マットソン（Mark Mattson）が統括した 2003 年の研究では，定期的に断食したマウスはカロリー制限をずっと続けたマウスに比べ，いくつかの指標でより健康的だった。例えば血中インスリン濃度と血糖値が低く，これはインスリン感受性が高く糖尿病のリスクが低いことを示している。

断食研究の 1 世紀

断食の精神的な効用は多くの宗教が古くから主張してきたことだが，身体的な利点が広く認識されるようになったのは 1900 年代初め，糖尿病や肥満，てんかんなど様々な疾患の治療に医師が断食を勧めるようになってからだ。

カロリー制限関連の研究が始まったのは 1930 年代で，コーネル大学の栄養学者マッケイ（Clive McCay）の発見がきっかけとなった。ラットを早くから厳しいカロリー制限のもとで育てると，自由に食べさせたラットよりも長生きし，年を取ってもがんなどの病気になりにくかったのだ。1945 年にはシカゴ大学の研究チームが，1 日おきに絶食したラットでも，先のマッケイの実験で毎日カロリー制限を続けたラットと同様に寿命が延びると報告し，カロリー制限と間欠的断食の研究が交わった。間欠的断食は寿命を延ばすだけでなく「死につながる疾患の発生を遅らせるようだ」とシカゴ大学のチームは論文に書いている。

その後数十年は，抗生物質の開発や冠動脈バイパス手術など，より影響力の大きな医学的進歩の陰に隠れて，抗老化食の研究は下火になった。しかし近年，高齢期における脳変性疾患のリスクを間欠的断食によって低下できるという考え方を，マットソンをはじめとする研究者が強く支持するようになった。彼らは，少なくとも齧歯類では間欠的断食が神経細胞を様々なストレス損傷から守ることを示した。1 日おきに絶食させたラットの場合，加齢に伴って細胞に損傷を引き起こす毒素に対して脳が強くなることを明らかにしたほか，後の研究で，間欠的断食が脳卒中のダメージを緩和し，パーキンソン病モデルマウスの運動障害を和らげ，遺伝子操作によってアルツハイマー病の症状を発現させたマウスの認知力低下を遅らせることを示した。彼自身，週末を除き朝食と昼食を抜く生活を続けている。「おかげで仕事がはかどる」という。55 歳のスリムなこの生物学者は共著を含め 700 本以上の論文を執筆してきた。

間欠的断食は細胞に分子レベルの損傷を防ぐ仕組みを常に働かせるよう仕向ける一種の軽いストレスとして機能しているのだとマットソンは考えている。例えば，たまに断食すると，細胞内に不適切な分子が生じるのを防ぐ「シャペロン・タンパク質」の濃度が上がる。また絶食中のマウスでは，ストレスを受けた神経細胞が死ぬのを防ぐ「脳由来神経栄養因子（BNDF）」の濃度が上がる。BNDFの低下は，うつ病からアルツハイマー病まですべての神経疾患と関連している（BNDFの低下が疾患の原因なのか結果なのかは不明）。このほか, 絶食はオートファジーを活発化する。これは細胞に備わった一種のゴミ処理機構で，損傷した分子を処分するものだ。そうした分子には，アルツハイマー病やパーキンソン病などの神経疾患と関連づけられてきたものが含まれている。

間欠的断食の主要効果のひとつは，身体のインスリン感受性を高めることだと思われる。インスリンは血糖値を調節しているホルモンで，インスリン感受性の低下は肥満を伴うことが多く，糖尿病や心臓病とも関連づけられてきた。動物も人間も，長生きの個体はインスリン濃度が非常に低い傾向がある。おそらく，細胞の感受性が高く，少量のインスリンで十分なためだろう。ソーク生物学研究所による最近の研究は，マウスに1日8時間は脂肪の豊富な餌を食べさせ，続く16時間は絶食させると，肥満せずインスリン濃度が危険水準に高まることもないことを示した。

プチ断食に連続的カロリー制限と同程度の利点がある（減量中もごちそうを多少食べられるかもしれない）ということで，これを試す人が増えてきたと，メリーランド大学の遺伝学教授マウント（Steve Mount）はいう。彼はプチ断食に関するヤフーのディスカッショングループの司会役を7年以上務めており，自らも2004年から週に3日の絶食を実行している。「間欠的断食は万能薬ではない。減量は難しいものだ」と断りながらも，間欠的断食とカロリー制限が同じ細胞内シグナル伝達経路を活性化しているという説は「理にかなっている」という。

まだ乏しい臨床データ

関心は高まっているものの，しっかりした臨床試験の例はまだ少なく，人間に対する長期的効果ははっきりしていない。それでも，ルイジアナ州の医師ジョンソン（James B. Johnson）は1957年のスペインの研究が参考になるとい

う。彼はこの研究結果を解析した論文を 2006 年に共著した。スペインの研究は高齢の男女 120 人を被験者とし，その半数に 1 日おきの絶食を 3 年間続けてもらった。この集団は期間中に平均 123 日を養護施設で暮らし，6 人が死亡した。これに対し絶食しなかった 60 人は平均 219 日を養護施設で送り，13 人が死亡した。

ジョンソンらは 2007 年，体重過多の喘息患者 9 人に 1 日おきにほぼ絶食する生活を 2 カ月続けてもらった結果，喘息の症状が急激に和らいだうえ，炎症の様々な兆候が治まったとする臨床試験結果を発表した。

だが一方では赤信号を発する研究もある。ブラジルで行われた 2011 年の研究は，間欠的断食を長期間続けると血糖値が上がるほか，細胞にダメージを与えかねない酸化性物質の組織内濃度が上昇することを示している。さらに，マットソンが共著した 2012 年の論文は，ラットを周期的に断食させたところ，なぜか心臓の組織が硬化し，血液を送り出す力が損なわれたと報告している。

また，減量の専門家のなかには，絶食時の空腹感や絶食後に大食いする恐れを挙げて懐疑的にみる向きもある。実際，アカゲザルでカロリー制限の効果を調べた最近の研究（長寿化の効果を認めなかったもの）は，食べ方を急に変える際には注意が必要であることを示している。

それでも進化の観点からいえば，1 日 3 度の食事というのは近代に発明された奇妙な習わしにすぎない。古代の祖先たちにとって食物供給は非常に不安定で，栄養不良や空腹はもちろん，飢餓に直面することも多かったろう。だがマットソンは，そうした進化上の圧力が働いた結果，学習と記憶に関連する脳領域を強化する一群の遺伝子が選択され，食物を見つけ出して生き延びられるようになったのだと考えている。この見方が正しいとすれば，間欠的断食はスマートになって生きるためのスマートな（賢い）方法なのかもしれない。

脳 & 老化

Blue Light Blues
青色光が奪う眠り

A Pain in the Brain
片頭痛に予防薬

Brain Food
地中海食と脳の健康

A Turn for the Worse
回転性めまい

Can We Stop Aging?
老化を止められるか？

Blue Light Blues
青色光が奪う眠り

パソコン画面を見ていると夜眠れなくなる理由と，その対策

F. ジャブル（SCIENTIFIC AMERICAN 編集部）

掲載：SCIENTIFIC AMERICAN November 2016, 日経サイエンス 2017 年 9 月号

About a decade ago Los Angeles–based software developer Lorna Herf decided to try her hand at oil painting. She and her husband, Michael, also a computer programmer, eventually installed bright fluorescent lights in their apartment's loft so that Lorna could paint at night and still have an accurate sense of what colors on the canvas would look like during the day. Late one evening Lorna descended to the living room, where computer screens were aglow. Now that she had become more attuned to differences in lighting, she noticed just how much the bright light from the computer screens clashed with the soft warmth of the incandescent bulbs that surrounded them. She remembers thinking the electronic screens were "like little windows of artificial daylight," spoiling the otherwise gentle ambience of the room.

The tech-savvy couple engineered a crafty solution to minimize the discrepancy. They wrote some code to change the number and wavelength of the photons emitted by their computer screens as a day progressed. The Herfs' goal was to mimic natural shifts in ambient light as closely as possible, transitioning from the bright, bluish-white light characteristic of morning and afternoon sunshine to a dim, orange glow in the evening.

At first, they simply intended to harmonize the lighting scheme in their home. But they soon began to sus-

Vocabulary

try one's hand at~ 試しに初めてやってみる

fluorescent light 蛍光灯

aglow 照り輝いて
attune to 慣れさせる, 適応させる

incandescent bulb 白熱電球

ambience 雰囲気

tech-savvy 技術通の
crafty 巧みな
discrepancy 不一致, 違和感
wavelength 波長
photon 光子

ambient light 周辺光

scheme 計画, 体制

pect that their new app, dubbed f.lux, might offer some health benefits as well. "After we'd been using it for a while, we started to notice it seemed easier to wind down at night," Lorna recalls, making it easier to fall asleep when they turned off their electronic devices. They are not the only ones who have appreciated the calming effect. Since the Herfs released the program for free in 2009, f.lux has been downloaded more than 20 million times.

By following their aesthetic taste, the Herfs had stumbled on a curious twist in the way the body controls how we sleep. Researchers have known for several decades that strong light of any kind can suppress melatonin, the hormone the brain produces at night to induce sleepiness. But more recent studies show that blue light suppresses melatonin more effectively than any other visible wavelength, potentially leaving people more alert when they would otherwise start feeling drowsy.

As it happens, smartphones, laptops and all kinds of electronic screens have become brighter and bluer over the past couple of decades because of the addition of powerful blue LEDs. During the day, when blue light is already naturally plentiful, a little extra exposure from electronic screens should not make much of a difference to anyone's physiology. The problem is that people are increasingly staring into bright screens long into the night.

Nearly everyone in a survey conducted by the National Sleep Foundation in 2011, for example, used a television, computer, cell phone or similar device within an hour of going to bed at least a few nights a week. In 2014 the same organization determined that 89 percent of adults and 75 percent of children in the U.S. have at least one electronic device in their bedroom, with a significant number of them sending or answering texts after they had initially fallen asleep. Motivated by such research, engi-

Vocabulary

app アプリ
wind down 緊張をほぐす, くつろぐ

appreciate 評価する, 認める

aesthetic 審美的な
stumble on 偶然に発見する

suppress 抑制する
melatonin メラトニン
induce 誘発する

alert 覚醒して
drowsy 眠い

as it happens あいにく, たまたま

LED 発光ダイオード (light emitting diode の頭字語)

physiology 生理
stare 見つめる

National Sleep Foundation 米国立睡眠財団

significant かなりの数の

2　脳&老化

neers and computer programmers are trying out various solutions to keep an already sleep-deprived population from losing more zzz's because of their electronic devices. The solutions range from tinted eyeglasses to naturalistic lighting systems for the home and office.

Vocabulary

sleep-deprived 睡眠不足の
zzz グーグー（いびきの音）, 眠り
tinted 色の入った

"If people can figure out ways to simulate changes in sunlight across the day, that would be perfect," says Christian Cajochen, head of the Center for Chronobiology at the University of Basel in Switzerland. "The ideal would be to have the same light throughout your home as outside of it." It remains to be seen how effective these remedies are, however, especially when compared with simply shutting the devices off.

chronobiology 時間生物学

remedy 矯正法, 治療法

TOO MUCH OF A GOOD THING

The light emanating from electronic devices was not always such a hindrance to restful sleep. The current state of affairs can be traced to the 1992 invention in Japan of the high-brightness blue LED. By combining the new blue LEDs with older green and red ones or coating blue LEDs with chemicals that reemit other wavelengths, technology manufacturers could generate full-spectrum white LED light for the first time. Because LEDs are much more energy-efficient than their fluorescent predecessors, they soon became ubiquitous in TVs, computer screens, tablets and certain e-readers, infusing homes and offices with much brighter blue light than ever before.

emanate 発する
hindrance 邪魔するもの
restful 安らかな

reemit 再放出する
full-spectrum フルスペクトルの, 全波長域にわたる

e-reader 電子書籍リーダー
infuse 持ち込む, 浸透させる

Researchers did not begin amassing concrete evidence that blue LEDs can disrupt sleep until about 15 years ago, but they have had a good idea of the probable mechanism for quite some time. Scientists had discovered back in the 1970s that a tiny brain region dubbed the supra-

amass 集める, 蓄積する
concrete 具体的な, 確かな

suprachiasmatic nucleus
▶ Technical Terms

Technical Terms　視交叉上核(**suprachiasmatic nucleus**)　脳の視床下部にある小さな神経核で, 体内時計をつかさどる中枢の役割を担っている。脳で左右の視神経が分岐して交差しているところを視交叉といい, そのすぐ上にあることからこの名がついた。

chiasmatic nucleus helps to control the body's sleep cycles, alertness, temperature and other daily fluctuations. Studies showed that the suprachiasmatic nucleus prompts the brain's pineal gland to produce melatonin every evening.

Earlier this century biologists uncovered exactly how this signaling process happens. As it turns out, the missing link was a previously unknown type of light-sensitive cell in the human eye, distinct from the familiar rods and cones that are responsible, respectively, for night and color vision. This third so-called photoreceptor tracks the amount of blue light in the environment and reports back to the suprachiasmatic nucleus. Thus, when there is a lot of blue light (as when the sun is overhead), this particular photoreceptor prompts the suprachiasmatic nucleus to tell the pineal gland not to make much melatonin, and so we stay awake. When the sun begins to set, however, the amount of blue light diminishes, triggering a surge in melatonin levels, prompting us to fall asleep.

Among the studies offering evidence that screens with blue LEDs might confuse the brain at night is a 2011 investigation by the University of Basel's Cajochen and his colleagues. In that work, volunteers exposed to an LED-backlit computer for five hours in the evening produced less melatonin, felt less tired, and performed better on tests of attention than those in front of a fluorescent-lit screen of the same size and brightness. Similarly, for subjects in a 2013 study led by Mariana Figueiro of the Rensselaer Polytechnic Institute, interacting with an iPad for just two hours in the evening was enough to prevent the typical nighttime rise of melatonin. And in a two-week trial at Brigham and Women's Hospital in Boston, published in 2014, volunteers who read on an iPad for four hours

Vocabulary

pineal gland
▶ Technical Terms

rod 桿体細胞
cone 錐体細胞
photoreceptor 光受容器

surge 高まり

confuse 混乱させる

attention 注意力

volunteer 被験者（実験に参加した有志）

Technical Terms　松果体（**pineal gland**）　脳にはホルモンを分泌する内分泌器がいくつかあり，松果体はそのひとつ。脳の中央奥深くにあり，直径1cmに満たない小さな器官だ。眠気を誘発するメラトニンというホルモンを分泌する。

before bed reported feeling less sleepy, took an average of 10 minutes longer to fall asleep and slept less deeply compared with those who read paper books at night. Cajochen and others have also shown that these effects are especially pronounced in teens and adolescents, for reasons that remain unclear.

pronounced 明白な, 顕著な
adolescent 若者

IN A NEW LIGHT

Given the accumulating evidence that artificial screens in general and blue lights in particular spoil sleep, scientists have begun investigating various remedies. Several studies have shown that wearing orange-tinted plastic goggles, which filter out the blue light emanating from electronic devices, helps to prevent melatonin suppression. Similar glasses are now commercially available for as little as $8 or as much as $100. A more expensive option is a so-called dynamic lighting system, which promises to re-create "the full range of natural daylight in an interior space" for hundreds to thousands of dollars depending on the size of one's home or office.

The most affordable countermeasures are computer programs such as f.lux. This past March, Apple introduced a function called Night Shift for the iPhone and iPad, which mimics f.lux in shifting the screen's emitted light "to the warm end of the spectrum" around sunset. So far no researchers have tested f.lux or Apple's Night Shift in a controlled study, but Figueiro says she is planning to conduct such experiments, and Michael Herf says he is collaborating with university scientists to examine the effects of f.lux in everyday environments outside the laboratory. "F.lux in my view is still a hypothesis," Herf adds. "We think it probably helps a lot of night owls, but we still need to support the anecdotes with data."

Researchers emphasize, however, that eliminating blue light is not a fail-safe solution. Even dim, orange screens make it tantalizingly easy to stay awake and read,

affordable 手ごろな
countermeasure 対抗策

controlled study 比較対照実験による研究

hypothesis 仮説
night owl 宵っ張りの人

anecdote 逸話, 風説, 事例

fail-safe 絶対安全な, まったく問題のない
tantalizingly 興味深くも

watch movies or play games at night, keeping your brain alert when it should be winding down. "It's as if you're completely in the dark, but you drink coffee," Figueiro explains. "It's still going to have an effect."

Ultimately the surest solution is electronic abstinence: shutting off all screens and bright lights for at least a few hours before bedtime. The inescapable fact is that humans evolved to rise and sleep with the sun. "Before we had all this technology, before electricity and artificial lighting, we would be awake in daylight, have a little bit of fire in the evening, and then sleep," says Debra Skene, a chronobiologist at the University of Surrey in England. Artificial light has been enormously beneficial over the centuries. But there are times, especially at the end of the day, when it can be too much of a good thing.

Vocabulary

abstinence 節制, 控えること

inescapable 不可避の, 逃れられない

chronobiologist 時間生物学者

10年ほど前，ロサンゼルスに住むソフトウエア開発者のローナ・ハーフ（Lorna Herf）は初めて油絵を描いてみることにした。やはりコンピュータープログラマーである夫のマイケル（Michael Herf）とともに自宅のロフトに明るい蛍光灯を取り付け，夜でもカンバス上の色合いが昼間と同じく正確に見えるようにして，制作にいそしんだ。ある夜遅く，ローナはロフトからパソコン画面があかあかとついている居間に下りて，あることに気づいた。油絵の練習を通じてライティングの差に敏感になっていたこともあって，画面からの青い光が周囲の白熱電球によるソフトで暖かな照明とぶつかっているのがはっきりわかった。パソコン画面がまるで「人工の日光が差し込む小さな窓」のように思えたのを覚えている。居間の柔和な雰囲気が，画面の近くだけは損なわれていた。

技術通の夫妻は，この違和感を最小化する巧妙な仕掛けを作った。パソコン画面が発する光の強さと波長を1日の進行に合わせて変えるプログラムを書いたのだ。周辺光に自然に生じる変化をできるだけ忠実にまねて，朝から午後の日光に特徴的な明るく青白い光から，夕暮れの薄暗いオレンジ色へと変えていく。

ハーフ夫妻は当初，自宅の照明環境を整えようとしただけだった。だが間もなく，「f.lux」と名づけたこのアプリがひょっとすると健康にも利点をもたらすのではないかと考えるようになった。「しばらく使っているうちに，夜にリラックスしやすくなるようだと気づいた」とローナは回想する。パソコンのスイッチを切ってからの寝つきがよくなった。この鎮静効果を認めているのは2人だけではない。夫妻が2009年にf.luxを無料公開して以降，このアプリは2000万回以上もダウンロードされている。

ハーフ夫妻は趣味の油絵を追求するうちにたまたま，人体が眠りを制御する方法に興味深いひとひねりを加える結果になった。強い光はどれも，脳が夜間に作り出して眠気を引き起こすメラトニンというホルモンを抑えることが数十年前から知られている。より最近の研究で，なかでも青色光がメラトニンを最も効果的に抑制することが示された。本来なら眠気を感じ始める時間なのに目が冴えたまま，という事態になりうる。

あいにく，スマートフォンやノートパソコンなど電子機器の画面は過去 20 年ほどでますます明るく青っぽくなった。高輝度の青色発光ダイオード（LED）が加わったおかげだ。自然環境に青色光があふれている昼間は，電子機器画面から多少の青色光が加わっても人間の生理にたいした違いは生じない。問題は，人々が夜遅くまで明るいスクリーンを見つめる機会が増えていることだ。

例えば米国立睡眠財団による 2011 年の調査では，調査対象者のほぼ全員が，就寝前 1 時間以内にテレビやパソコン，携帯電話などの機器を使う日が週に数日はあると回答した。同財団は 2014 年，米国の成人の 89％と未成年者の 75％が自分の寝室に電子機器を少なくとも 1 つ保有しており，かなりの人が，いったん眠りに落ちた後に再び起きてメールを送受信していると結論づけた。技術者やコンピュータープログラマーはこうした調査に刺激され，すでに睡眠不足にある人々が電子機器のせいで眠りをさらに奪われるのを避ける様々な方策を試している。色つきメガネから，住居・オフィス向けの自然調光システムまで，対策は様々だ。

「1 日を通じた日光の変化を模擬する方法を開発できれば，それが最善だろう」と，スイスのバーゼル大学で時間生物学センターの所長を務めているカヨーヒェン（Christian Cajochen）はいう。「室内の光をすべて戸外と同じにするのが理想だ」。ただ，効果のほどは今後の研究課題。単に電子機器のスイッチを切れば十分かもしれない。

過ぎたるは及ばざるがごとし

電子機器の発する光が常に安眠を妨げてきたわけではない。現在の状況は，1992 年に日本が高輝度青色 LED を発明したことにさかのぼる。これを既存の緑と赤の LED と組み合わせるか，光を別波長に変えて放射する化学物質を青色 LED にコーティングすることで，フルスペクトルの白色 LED が初めて作られた。LED は蛍光灯よりもエネルギー効率がよいので，すぐにテレビやパソコン画面，タブレット端末，一部の電子書籍リーダーなどに広く採用され，住居やオフィスの空間にかつてない明るい青色光成分が浸透した。

青色LEDが睡眠を妨げうる証拠が集まり始めたのは15年ほど前からだが，青色光に予想される効果についてはかなり前からよくわかっていた。1970年代には，脳の「視交叉上核」という小さな領域が睡眠周期と覚醒，体温などの日次変動の制御に関与していることが発見された。視交叉上核は脳にある内分泌器「松果体」を刺激し，メラトニンを毎晩作り出す。

21世紀に入ると，このプロセスがどのように進むのかが解明された。眼のなかに，よく知られた桿体細胞と錐体細胞（それぞれ夜間の視覚と色覚を担っている）とは別の光感受性細胞が存在することが明らかになった。この第3の光受容器は周辺環境に存在する青色光の量を追跡し，それを視交叉上核に伝えている。青色光が豊富なとき（太陽が出ているときなど）には，この光受容器が視交叉上核にその情報を連絡して，メラトニンをあまり作らないよう松果体に指示を出している。だから私たちは昼の間ずっと目覚めている。日が暮れると青色光の量が減り，メラトニンの量が増えて眠気を誘うのだ。

青色LEDの画面が夜間に脳を混乱させている可能性を示した研究例に，カヨーヒェンらバーゼル大学のチームが行った2011年の実験がある。夜にLEDバックライトのパソコン画面を5時間見た被験者は，同じ明るさとサイズの蛍光灯照明画面の前に座っていた人に比べ，メラトニンの生成量が少なく，疲れを感じず，注意力テストの成績が高かった。同様に，レンセラー工科大学のフィギュイロウ（Mariana Figueiro）が率いた2013年の研究では，夜にiPadをたった2時間操作しただけで，典型的なメラトニンの上昇が起こらなくなった。そして2014年に発表されたブリガム・アンド・ウィメンズ病院における2週間の試験では，就寝前にiPadで4時間読書した被験者は紙の本を読んだ人と比較して，眠気を感じず，眠りに落ちるのが平均で10分遅くなり，眠りが浅くなったと感じた。カヨーヒェンらはまた，これらの効果が特に10代の青少年と若者に強く生じることを示した。理由はまだわかっていない。

最も確実な解決策は…

人工光，特に青色光が睡眠を妨げる証拠が集積したことから，様々な矯正法が探られてきた。オレンジ色のプラスチック製ゴーグルを着用して電子機器からの青色光を遮るとメラトニンの抑制を防ぐ効果があることが複数の研究か

ら示された。この種の遮光メガネは 8 ドルから 100 ドルまで，様々な価格のものが売られている。さらに高価ではあるが，「ダイナミック照明システム」という手段もある。「フルレンジの自然光を室内空間に再現する」とされるもので，住居やオフィスの大きさによって数百ドルから数千ドルかかる。

最も手ごろな対策は f.lux のようなコンピュータープログラムだ。アップルは 2016 年 3 月，iPhone と iPad に「Night Shift（ナイトシフト）」という機能を導入した。f.lux のように，画面が発する光を「スペクトルの暖色側へ」シフトさせ，日没のころと似た状態にする。これまでのところ f.lux やナイトシフトを比較対照実験で調べた研究はないが，フィギュイロウがその種の厳密な実験を計画しているほか，マイケル・ハーフは大学の科学者と共同で f.lux の効果を実験室外の日常環境で調べている。「f.lux はまだ仮説のレベルだ。宵っ張りの人たちに役に立つとは思われるが，それを裏づけるデータが必要だ」という。

ただ，青色光を排除すれば万全というわけではないと研究者たちは強調する。オレンジ色の薄暗い画面であっても，夜に目覚めたまま画面を読み，映画鑑賞やゲームをしたい気になり，本来なら脳が静まるべき時間なのに頭が冴えたままになる。「真っ暗闇のなかでコーヒーを飲んで頑張っているようなものだ」とフィギュイロウは説明する。「その影響は免れないだろう」。

結局のところ，最も確実な解決策は電子機器の使用を控えることだ。就寝の少なくとも 2 〜 3 時間前にはすべての画面と明るい照明のスイッチを切ること。人間が太陽に合わせて起き眠るよう進化してきたという事実は動かせない。「これらの技術がなかったころ，電気と人工照明がなかったころ，人間は日中に起きて活動し，日没後は少し火をたいて，後はさっさと眠っていたのだ」と英サリー大学の時間生物学者スキーン（Debra Skene）はいう。人工光は過去何世紀にもわたって多大な便益をもたらしてきたが，いかに有用なものでも，過ぎたるは及ばざるがごとしとなる場合がある。

A Pain in the Brain
片頭痛に予防薬

三叉神経の過剰反応に着目した研究が，初の発作予防薬につながった

D. ヌーナン（サイエンスライター）

掲載：SCIENTIFIC AMERICAN December 2015，日経サイエンス 2016 年 6 月号

The 63-year-old chief executive couldn't do his job. He had been crippled by migraine headaches throughout his adult life and was in the middle of a new string of attacks. "I have but a little moment in the morning in which I can either read, write or think," he wrote to a friend. After that, he had to shut himself up in a dark room until night. So President Thomas Jefferson, in the early spring of 1807, during his second term in office, was incapacitated every afternoon by the most common neurological disability in the world.

The co-author of the Declaration of Independence never vanquished what he called his "periodical head-ach," although his attacks appear to have lessened after 1808. Two centuries later 36 million American migraine sufferers grapple with the pain the president felt. Like Jefferson, who often treated himself with a concoction brewed from tree bark that contained quinine, they try different therapies, ranging from heart drugs to yoga to herbal remedies. Their quest goes on because modern medicine, repeatedly baffled in attempts to find the cause of migraine, has struggled to provide reliable relief.

Now a new chapter in the long and often curious history of migraine is being written. Neurologists believe they have identified a hypersensitive nerve system

Vocabulary

migraine 片頭痛

shut up 閉じ込める

incapacitate 無能にする

Declaration of Independence アメリカ独立宣言
vanquish 征服する

grapple with 取っ組み合う
concoction 調合薬

quinine キニーネ

quest 探求
be baffled in 〜に失敗する

hypersensitive 過敏な

that triggers the pain and are in the final stages of testing medicines that soothe its overly active cells. These are the first ever drugs specifically designed to prevent the crippling headaches before they start, and they could be approved by the U.S. Food and Drug Administration next year. If they deliver on the promise they have shown in studies conducted so far, which have involved around 1,300 patients, millions of headaches may never happen.

"It completely changes the paradigm of how we treat migraine," says David Dodick, a neurologist at the Mayo Clinic's campus in Arizona and president of the International Headache Society. Whereas there are migraine-specific drugs that do a good job stopping attacks after they start, the holy grail for both patients and doctors has been prevention.

Migraine attacks, which affect almost 730 million people worldwide, typically last from four to 72 hours. Most sufferers have sporadic migraines and are laid low during 14 or fewer days a month. Those with a chronic form—almost 8 percent of the migraine population—suffer 15 or more monthly "headache days." Attacks are often preceded by fatigue, mood changes, nausea and other symptoms. About 30 percent of migraine patients experience visual disturbances, called auras, before headaches hit. The total economic burden of migraine in the U.S., including direct medical costs and indirect costs such as lost workdays, is estimated at $17 billion annually.

In the 5,000 years since migraine symptoms were first described in Babylonian documents, treatments have reflected both our evolving grasp and our almost comical ignorance of the condition. Bloodletting, trepanation and cauterization of the shaved scalp with a red-hot iron bar were common treatments during the Greco-Roman period. The nadir of misguided remedies was probably reached in the 10th century A.D., when the otherwise discerning ophthalmologist Ali ibn Isa recommended binding a dead mole to the head. In the 19th century medical electricity had become all the rage, and migraine patients were routinely jolted with a variety of inventions, including the hydroelectric bath, which was basically an electrified tub of water.

By the early 20th century clinicians turned their attention to the role of the blood vessels, inspired in part by observations of strong pulsing of the temporal arteries in migraine patients, as well as patients' descriptions of throbbing pain and the relief they got from compression of the carotid arteries. For decades to come, migraine pain would be blamed primarily on the dilation of blood vessels (vasodilation) in the brain.

That idea was reinforced in the late 1930s with the publication of a paper on the use of ergotamine tartrate, an alkaloid that was known to constrict blood vessels. Despite an array of side effects, among them vomiting and drug dependence, it did stop attacks in a number of patients.

But if vasodilation was part of the puzzle, it was not the only thing going on in the brains of migraine sufferers, as the next wave of treatments suggested. In the 1970s cardiac patients who also had migraines started telling their doctors that the beta blockers they were taking to slow rapid heartbeats also reduced the frequency of their attacks. Migraine sufferers taking medicines for epilepsy and depression, and others receiving cosmetic Botox injections, also reported relief. So headache specialists began prescribing these "borrowed" drugs for migraines. Five of the medications eventually were approved by the FDA for the condition. Unfortunately, it is still not known exactly how the adopted drugs (which are effective in only about 45 percent of cases and come with an array of side effects) help migraines. Dodick says they may act at vari-

Vocabulary

ophthalmologist 眼科医
become all the rage 大流行に至る
hydroelectric bath 電気水浴

temporal arteries 側頭動脈

throbbing ずきずきする
carotid artery 頸動脈
dilation 拡張, 膨張
vasodilation 血管拡張

ergotamine tartrate 酒石酸エルゴタミン
alkaloid アルカロイド
constrict 収縮させる
vomiting 嘔吐

beta blockers β遮断薬

epilepsy てんかん
depression うつ病
cosmetic Botox injections しわ取り美容のボトックス注射

adopt 借用する

ous levels of the brain and brain stem to reduce excitability of the cortex and pain-transmission pathways.

The first migraine-specific drugs, the triptans, were introduced in the 1990s. Richard Lipton, director of the Montefiore Headache Center in New York City, says triptans were developed in response to the older idea that the dilation of blood vessels is the primary cause of migraine; triptans were supposed to inhibit it. Ironically, subsequent drug studies show that they actually disrupt the transmission of pain signals in the brain and that constricting blood vessels is not essential. "But they work anyway," Lipton says. A survey of 133 detailed triptan studies found that they relieved headache within two hours in 42 to 76 percent of patients. People take them to stop attacks after they start, and they have become a reliable frontline treatment for millions.

What triptans cannot do—and what Peter Goadsby, director of the Headache Center at the University of California, San Francisco, has dreamed about doing for more than 30 years—is prevent migraine attacks from happening in the first place. In the 1980s, in pursuit of this goal, Goadsby focused on the trigeminal nerve system, long known to be the brain's primary pain pathway. It was there, he suspected, that migraine did its dirty work. Studies in animals indicated that in branches of the nerve that exit from the back of the brain and wrap around various parts of the face and head, overactive cells would respond to typically benign lights, sounds and smells by releasing chemicals that transmit pain signals and cause migraine. The heightened sensitivity of these cells may be inherited; 80 percent of migraine sufferers have a family history of the disorder.

Vocabulary
brain stem 脳幹

triptan トリプタン

inhibit 阻害する

frontline 第一線の

trigeminal nerve system
▶ Technical Terms

wrap around 巻きつく

benign 穏やかな

family history 家族歴

Technical Terms
三叉神経系(**trigeminal nerve system**)　脳神経(脳から直接出ている末梢神経)は12対あり、三叉神経系はその1つ。脳神経のなかでは最大だ。眼神経と上顎神経、下顎神経の3つに分かれ、主に頭部の感覚と運動制御を担っている。

Goadsby co-authored his first paper on the subject in 1988, and other researchers, including Dodick, joined the effort. Their goal was to find a way to block the pain signals. One of the chemicals found in high levels in the blood of people experiencing migraine is calcitonin gene-related peptide (CGRP), a neurotransmitter that is released from one nerve cell and activates the next one in a nerve tract during an attack. Zeroing in on CGRP and interfering with it was hard. It was difficult to find a molecule that worked on that neurotransmitter and left other essential chemicals alone.

As biotech engineers' ability to control and design proteins improved, several pharmaceutical companies developed migraine-fighting monoclonal antibodies. These designer proteins bind tightly to CGRP molecules or their receptors on trigeminal nerve cells, preventing cell activation. The new drugs are "like precision-guided missiles," Dodick says. "They go straight to their targets."

It is that specificity, and the fact that scientists actually know how the drugs work, that has Dodick, Goadsby and others excited. In two placebo-controlled trials with a total of 380 people who had severe migraines up to 14 days per month, a single dose of a CGRP drug decreased headache days by more than 60 percent (63 percent in one study and 66 percent in the other). In addition, in the first study, 16 percent of the patients remained totally migraine-free 12 weeks into the 24-week trial. Larger clinical trials to confirm those findings are currently under

Vocabulary

calcitonin gene-related peptide カルシトニン遺伝子関連ペプチド
neurotransmitter
▶ Technical Terms
zero in on ～に的を絞る

leave ~ alone ～をそのままにしておく

monoclonal antibody
▶ Technical Terms

precision-guided missile 精密誘導ミサイル

specificity 特異性

placebo-controlled trial プラセボ対照試験

Technical Terms

神経伝達物質(**neurotransmitter**)　ニューロン(神経細胞)の末端から放出され、隣接ニューロンに伝わってそれを興奮または抑制する物質。ニューロンどうしの接続部はシナプスという狭い隙間になっており、神経伝達物質はこのシナプス間隙を拡散して、受け手側ニューロンの受容体に結合する。様々な種類の神経伝達物質が知られている。
モノクローナル抗体(**monoclonal antibody**)　抗原を特徴づける分子領域は複数あるが、そのうち1つだけを認識して結合する抗体のこと。特異的に作用する均一な分子なので、各種の診断・治療薬として優れた働きを示す。

way. So far the CGRP drugs work better at prevention than any of the borrowed heart or epilepsy drugs and have far fewer side effects. They are given to patients in a single monthly injection.

Migraine specialists are also exploring other treatments, including forehead and eyelid surgery to decompress branches of the trigeminal nerve, as well as transcranial magnetic stimulation (TMS), a noninvasive way of altering nerve cell activity.

Lipton says he has had some good results with TMS. He has also referred patients for surgical interventions but says the experience "has been disappointing," and he is not recommending it. For his part, Goadsby views surgeries and high-tech efforts as a kind of desperation: "They strike me as a cry for help. If we better understood migraine, we'd know better what to do."

Even though the cause now appears rooted in the trigeminal nerve system, the origin of its overactive cells is still a mystery, Goadsby says. "What's the nature of what you inherit when you inherit migraine?" he asks. "Why you, and why not me?" If researchers untangle the genetics of migraine, Jefferson's "periodical head-ach" may loosen its painful modern grip.

Vocabulary

forehead 額
eyelid まぶた
decompress 減圧する
transcranial magnetic stimulation
▶ Technical Terms

desperation 自暴自棄

untangle 読み解く

Technical Terms 経頭蓋磁気刺激(**transcranial magnetic stimulation**)　電磁石で発生した変動磁場によって脳内に誘導電流を生み出し、これによって脳内のニューロンを興奮させる方法。神経科学の研究のほか、一部の神経疾患の診断に使われている。治療目的の応用はまだ少ない。

その63歳の政府要人は仕事ができなかった。長年にわたり片頭痛を抱え，またも新たな発作に襲われていた。友人あての手紙に「その日の午前中，読むことも書くことも，考えることもほとんどできなかった」と書いている。当日は日暮れまで暗い部屋にこもって休まざるをえなかった。1807年早春，2期目の大統領職にあったジェファーソン（Thomas Jefferson）はこのように，片頭痛という最も一般的な神経疾患のせいで毎日の午後は就業不能だった。

翌年以降はましになったようだが，このアメリカ独立宣言の起草者が「定期的な頭痛」と呼んだものから解放されることはついぞなかった。200年後の現在，3600万人の米国人がジェファーソン大統領と同じ頭痛と格闘している。そして彼がしばしば樹皮を煎じてキニーネを含む調合薬を作って飲んだように，現代の患者も心臓病薬からヨガ，ハーブ療法まで様々な方法を試している。手探りが続いているのは，片頭痛の原因解明を図る試みが失敗の繰り返しに終わり，現代医学が確実な救援策の提供にいまだに苦闘しているからだ。

だが，片頭痛をめぐる長くて時に奇妙な歴史に，いま新たな一章が書き加えられつつある。片頭痛の引き金とみられる過敏な神経系が特定され，これを鎮める薬が臨床試験の最終段階に入った。片頭痛が始まる前に発作を防ぐことを狙った初めての薬であり，2016年中に米食品医薬品局（FDA）から認可される可能性がある。約1300人を対象にしたこれまでの試験では有望な結果が出ており，何百万何千万もの人々が片頭痛からついに解放されるかもしれない。

「片頭痛の治療が根本的に変わる」というのは，メイヨークリニック・アリゾナキャンパスの神経科医で国際頭痛学会（HIS）の会長を務めているドディック（David Dodick）だ。発作開始後に痛みを抑える片頭痛薬はあるものの，患者と医師が念願としてきたのは発作を未然に防ぐ薬だ。

片頭痛患者は世界で7億3000万人近い。発作は通常，4時間から72時間続く。患者の多くは"頭痛日"が1カ月に14日以下だが（孤発性片頭痛），8%近くは月に15日以上の慢性片頭痛だ。発作の前に疲労感や気分変調，吐き気などの症状が表れることが多い。患者の約30%は頭痛の前に「閃輝暗点」という視覚の乱れを経験する。米国における片頭痛による経済損失は，直接の医療費と欠勤な

どによる間接的な損失を合わせて，年間 170 億ドルと推定されている。

　片頭痛に関する最古の記述は 5000 年前のバビロニアの文書に見られる。以来，その治療は片頭痛に関する人々の理解の進展と滑稽なまでの無知を反映してきた。ギリシャ・ローマ時代には放血と頭部穿孔，毛を剃った頭蓋に焼きごてを当てて焼灼するのが一般的な治療法だった。見当違い療法の最たるものは 10 世紀，他の点では見識豊かな眼科医だったアリ・イブン・イシャ（Ali ibn Isa）が推奨した方法で，頭にモグラの死骸をくくりつけるというもの。19 世紀には電気療法が大流行し，片頭痛患者も様々な発明品で電気刺激を受けた。ふろの湯に電気を通してつかる電気水浴などだ。

　20 世紀初頭になると，医師たちは血管の役割に注意を向けた。片頭痛患者の側頭動脈に強い拍動が見られるほか，この頭痛が「ずきずきする痛み」であること，頸動脈を圧迫すると痛みが和らぐとの患者の声がきっかけだった。その後数十年，片頭痛は主に脳の血管拡張が原因であるとされた。

　この考え方を裏づける研究が 1930 年代後半に発表された。血管を収縮させるアルカロイドとして知られる酒石酸エルゴタミンに関するもので，吐き気や依存性など一連の副作用があるものの，この物質は一部の患者の発作を確かに止めた。

　だが，血管拡張が関与しているとしても，それだけではないことが 1970 年代に示唆された。片頭痛の心臓病患者が，動悸を抑えるためにβ遮断薬を服用すると片頭痛の発作も減ると報告し始めたのだ。抗てんかん薬や抗うつ薬を服用している患者や，しわ取り美容のボトックス注射を受けている人も同様の改善を報告した。そこで頭痛専門医はこうした"適応外"の薬を片頭痛に処方し始めた。そのうち 5 つは最終的に片頭痛向けに正式に認可された。だがあいにく，これらの薬（片頭痛症例の約 45％に効果があるのみで，様々な副作用もある）がどうして効くのか正確にはわかっていない。脳と脳幹の様々なレベルに作用して，皮質と痛み伝達神経経路の興奮性を下げているのかもしれないとドディックはいう。

初の片頭痛専門薬「トリプタン」は1990年代に実用化した。この薬は血管拡張が片頭痛の主因であるとする古い考え方のもとで開発されたと，モンテフィオーレ医療センター（ニューヨーク）の所長リプトン（Richard Lipton）はいう。つまり，トリプタンは血管拡張を阻害するのだと考えられた。だが皮肉なことに，トリプタンは実は脳での痛み信号伝達を阻害していて，血管の収縮は必要ではないことが後の研究で判明した。「だが，この薬はともかく効いた」とリプトンはいう。トリプタンに関する133件の研究を調べた結果，患者の42〜76％で2時間以内に頭痛が治まっていた。生じた頭痛を止める薬として，何百万人もの片頭痛患者が服用する第一選択の薬となった。

だが，発作そのものの発生を防ぐことはトリプタンにはできない。カリフォルニア大学サンフランシスコ校頭痛センター所長のゴーズビー（Peter Goadsby）は30年以上にわたって片頭痛を未然に防ぐことを夢見てきた。このため彼は1980年代，主要な疼痛信号伝達経路として知られる三叉神経系に注目した。片頭痛が悪さをしているのはここだろうとにらんで動物実験を重ね，三叉神経のうち後頭部の脳に発して顔面と頭部の様々な部分をカバーしている枝において一部の細胞が光や音，においなどの穏やかな刺激に過剰反応し，痛み信号を伝える物質を放出することで片頭痛を引き起こしているらしいことを見いだした。これらの細胞の過敏さは遺伝的に受け継がれるようで，片頭痛患者の80％はこの病気の家族歴がある。

ゴーズビーはこのテーマを扱った初の論文を1988年に共著し，ドディックら他の専門家も研究に加わった。目標は疼痛信号をブロックする方法を見つけることだ。片頭痛の最中に血中濃度が高まる物質に「カルシトニン遺伝子関連ペプチド（CGRP）」がある。発作の間に神経細胞から放出されて隣の神経細胞を活性化している神経伝達物質だ。だが，CGRPに的を絞って阻害するのは難しい。CGRPに作用し，他の重要な神経伝達物質には影響を与えないような分子を見つけるのは困難だった。

その後バイオ技術によってタンパク質を制御・設計する能力が高まり，医薬品メーカー数社が片頭痛と闘うモノクローナル抗体を開発した。CGRP分子あるいは三叉神経系の神経細胞にある受容体に結合し，細胞の活性化を防ぐ。「精

密誘導ミサイルのように標的に直行する」とドディックはいう。

　　この特異性，そして薬がどのように作用するかを科学者が実際に把握できている点が，従来との違いだ。1カ月に最大で14日までひどい片頭痛が生じる合計380人を対象に行われた2件のプラセボ対照試験で，CGRP阻害薬を1回投与することで頭痛日が60％以上も減った（一方の試験では63％減，他方は66％減）。さらに，最初に行われたほうの試験では，16％の患者が24週間の試験期間中のうち12週間は頭痛がまったく起こらなかった。現在，被験者数を増やした大規模な臨床試験で確認が進められている。これまでのところ，心臓病薬などの転用薬よりも優れた予防効果を発揮しており，副作用もはるかに少ない。月に1回，注射で投与する。

　　別の治療法も探求されている。額やまぶたを手術して三叉神経の枝を減圧する方法や，経頭蓋磁気刺激法（TMS）という非侵襲的な手段で神経細胞の活動を調整する方法などだ。

　　リプトンは経頭蓋磁気刺激で良好な結果をいくつか得ているという。外科手術を選択した患者もあったが，結果は「期待外れ」で，現在は推奨していない。一方のゴーズビーはこうした手術やハイテクに頼るのは一種の破れかぶれだとみる。「私には助けを求める悲鳴に聞こえる。片頭痛に関する理解が改善すれば，何をすべきかがもっとよく見えてくるだろう」。

　　片頭痛の原因が三叉神経系にあるらしいとわかった現在も，過敏な神経細胞がそもそもなぜ生じるのかは依然として謎のままだとゴーズビーはいう。「遺伝で片頭痛を受け継いだとき，その受け継がれたものの正体とはいったい何なのか？」と彼は問う。「そして，片頭痛を患う人とそうでない人がいるのはなぜなのか？」　片頭痛の遺伝学を読み解けば，ジェファーソンの「定期的な頭痛」が現代人に及ぼす痛みも解けてくるだろう。

Brain Food
地中海食と脳の健康

記憶力低下を遅らせるようで，年を取ってからでも効果が期待できる

D. F. マロン（SCIENTIFIC AMERICAN 編集部）

掲載：SCIENTIFIC AMERICAN September 2015, 日経サイエンス 2016 年 3 月号

Whenever the fictional character Popeye the Sailor Man managed to down a can of spinach, the results were almost instantaneous: he gained superhuman strength. Devouring any solid object similarly did the trick for one of the X-Men. As we age and begin to struggle with memory problems, many of us would love to reach for an edible mental fix. Sadly, such supernatural effects remain fantastical. Yet making the right food choices may well yield more modest gains.

A growing body of evidence suggests that adopting the Mediterranean diet, or one much like it, can help slow memory loss as people age. The diet's hallmarks include lots of fruits and vegetables and whole grains (as opposed to ultrarefined ones) and a moderate intake of fish, poultry and red wine. Dining mainly on single ingredients, such as pumpkin seeds or blueberries, however, will not do the trick.

What is more, this diet approach appears to reap brain benefits even when adopted later in life—sometimes aiding cognition in as little as two years. "You will not be Superman or Superwoman," says Miguel A. Martínez González, chair of the department of preventive medicine at the University of Navarra in Barcelona. "You can keep your cognitive abilities or even improve them

Vocabulary

down 平らげる

devour むさぼり食う

edible 食べられる
fix 修理, 治療

Mediterranean diet 地中海式ダイエット, 地中海食
hallmark 特徴
whole grains 全粒穀物

poultry 鶏肉
ingredient 成分, 料理の材料

reap 収穫を得る, 刈り取る肉

cognitive ability 認知能力

slightly, but diet is not magic." Those small gains, however, can be meaningful in day-to-day life.

Vocabulary

day-to-day 日々の

FROM FORK TO BRAIN

Scientists long believed that altering diet could not improve memory. But evidence to the contrary started to emerge about 10 years ago. For example, Nikolaos Scarmeas of Columbia University and his colleagues collected information about the dietary habits and health status of about 2,000 Medicare-eligible New Yorkers—typically in their mid-70s—over the course of four years on average. In 2006 the investigators reported that tighter adherence to a Mediterranean diet, which had previously been linked to a lower risk of cardiovascular disease, was associated with slower cognitive decline and a lower likelihood of acquiring Alzheimer's disease. Because the researchers merely observed dietary patterns and did not control them—as would be the case in a clinical trial—doubts lingered, however. It was still possible that the apparent brain benefit was the result of chance or some other trait common to folks who consistently follow a Mediterranean diet in the U.S., such as educational achievement or particular life choices.

Medicare メディケア（米国の高齢者・障害者向け公的医療保険制度）
eligible 有資格者

cardiovascular disease 心血管疾患

control 対照群と比較する
clinical trial 臨床試験

chance 偶然
trait 形質

Seven years later researchers pinned down some answers. In 2013 Martínez González and his colleagues published findings on their massive PREDIMED study, an experiment that included almost 7,500 people in Spain. (PREDIMED stands for Prevention with Mediterranean Diet.) The investigators randomly assigned study subjects to one of two experimental groups. In the first, participants followed the Mediterranean diet with an additional helping of mixed nuts; in the second, they also adhered to the Mediterranean diet but were given additional extra virgin olive oil. (Researchers felt that providing extra nuts and oils at no cost to participants would guarantee that certain healthy fats were eaten in quantities large enough to have measurable effects on the study's outcomes.) The

experimental group 実験群

helping 一盛り

control group, against which the results of the experimental groups would be compared, was instructed generally on how to lose weight. Its members were given advice on eating vegetables, meat and high-fat dairy products that jibed with the Mediterranean diet, but they were discouraged from using olive oil for cooking and from consuming nuts.

As expected, the results showed that either of the experimental Mediterranean diet options led to significantly better cardiovascular outcomes. But when the scientists tested cognition in a subset of study members, they also discovered that individuals in either of the Mediterranean diet groups performed better than the weight-instruction group in a battery of widely accepted cognitive tests. "This is surprising, of course," Martínez González says.

As intriguing as these findings are, they are still not conclusive; the researchers had not gathered any cognitive information at the beginning of the study. Therefore, the possibility remains that there was something different between the two experimental groups and the control group—beyond their diet interventions—that could account for the findings.

Martínez González sought to quiet such criticisms with a new study his team published in July in *JAMA Internal Medicine*. Drawing from a group of more than 300 participants who were also part of PREDIMED but at a specific site with more financial resources, the researchers conducted baseline cognitive measurements and compared them with that same group's results four years later. On average, people were 67 years old at the start of the study. The newest findings, Martínez González says, are consistent with what he found in his earlier studies. These results are also not definitive, however, because this substudy was relatively small. Yet, he notes, it is the first time scientists have seen improvements in cognitive function from a randomized trial of the Mediterranean diet.

Can Americans, whose standard diet and way of life are often substantially different from that of adults living in Spain, benefit from the approach? That remains to be seen. The normal diet of the people in the study's control group was still closer to a Mediterranean diet than that of most Americans, so they already had years of relatively healthy eating under their belts, which could have helped their overall health. But Martínez González believes that the diet might provide even greater benefits for Americans because they have so much more room for improvement. Still, nutrition expert Martha Morris of Rush University says, only a randomized trial in the U.S. can truly answer the question—something she hopes to spearhead in the coming years.

BEYOND DIET

Proving that a particular cuisine affects cognitive health is one thing. Getting a lot of Americans to eat more fruits, vegetables, fish and olive oil is another matter altogether. Two major obstacles are cost and ingrained habits. For PREDIMED, study participants were supplied with expensive extra virgin olive oil and told how to prepare meals. "To transfer this knowledge to the American population, you can't just show them food items," Martínez González says. "You have to show them how to shop for them, cook with them and prepare them to keep all the nutrients in line with the traditional Mediterranean diet." The first step in the right direction, he says, would be for Americans to slash their consumption of red meats and use poultry instead. But that still leaves a lot of other steps to go before they are eating a Mediterranean diet.

Adhering to the exact diet laid out in PREDIMED may not be the only way to gain cognitive benefits from

Vocabulary

randomized trial ランダム化比較試験
▶ 111 ページ **Technical Terms**

eat under one's belt 食べて胃に収める

spearhead 陣頭指揮する

ingrained 深く染みこんだ

red meat 赤身肉, 獣肉

food. In February, Morris and her colleagues published online a study recommending a modified diet largely consistent with the Mediterranean diet but one cheaper to adopt in the U.S. Morris's so-called MIND diet emphasizes green, leafy plant and whole grain consumption. Its staples include two veggie servings a day, two berry servings a week and, instead of the almost daily fish consumption required in the Mediterranean diet, fish only once a week.

Morris found that even moderate adherence to the MIND diet for an average of 4.5 years appeared to reduce Alzheimer's risk compared with the Mediterranean and another diet. She and her colleagues judged that outcome by counting the number of cases of clinically diagnosed Alzheimer's among each group during the study period. Better yet, the MIND diet may be more achievable for the average person's wallet and for American culture. In the bigger picture, this finding suggests that "people improving their diet can make a difference for their memory," says Francine Grodstein, a professor focusing on healthy aging at Brigham and Women's Hospital in Boston and Harvard Medical School, who was not involved with the work.

Why certain food choices might help the brain function better remains unclear. Perhaps these regimens' known cardiovascular benefits, which promote a good flow of blood and oxygen to the brain, are key. But other factors may be at work. Of course, questions about when these dietary changes need to happen or how diet stacks up against other factors, such as physical activity, sleep patterns and genetics also remain unanswered.

Recently some researchers have begun broadening their focus beyond food alone. In the European Union, a multicountry randomized trial beginning this year is designed to provide further insights into how diet, exercise and better control of blood pressure could work together

Vocabulary

consistent with 〜と一致した，変わらない
staple 主成分，中心的要素

stack up against と比べて〜である

to promote brain health. (Hypertension is a leading cause of stroke, which can seriously harm mental processing.) Although the study will not allow scientists to pinpoint which factor offers the greatest benefit, it should give them a better understanding of how significant a role life changes can play.

There is reason to be hopeful. A pilot study published in June in the *Lancet* found that making changes in diet and habits later in life can slow the course of cognitive decline. Scandinavian researchers divided a group of 1,260 people in Finland either to receive standard nutrition and diet advice or to follow a specified exercise plan and eat a modified Mediterranean diet—all while their blood pressure and other health indicators were monitored and, if necessary, treated. Subjects in the experimental group ended up doing significantly better on standard tests of cognition. "We could really see that [the intervention] can protect against or at least delay cognitive impairments," says lead study author Miia Kivipelto, director of research and education at the geriatric clinic at the Karolinska Institute in Stockholm. Unexpectedly, she says, those changes were visible within just two years. And best of all, superpowers are not required.

Vocabulary

hypertension 高血圧
stroke 脳卒中

geriatric 老人病学の, 老年医学の

ポパイがホウレンソウの缶詰を平らげると，いつも決まって効果てきめん，すぐに超人的な力が出る。X-メンにも同様のトリックを使うミュータントがいた。年を取って物忘れしやすくなり始めた多くの人も，これを食べるとたちまち記憶力回復——となればよいのだが，残念ながらそんな超自然的効果はおとぎ話の世界。だが，適切な食べ物を選ぶと，それなりの改善は期待できそうだ。

地中海式ダイエット（地中海食）やそれに類した食事が加齢に伴う記憶力低下を遅らせるのに有効であることが，多くの研究から示唆されている。地中海食はたっぷりの野菜と果物，精白していない全粒穀物を中心に，適量の魚や鶏肉，赤ワインをとるのが特徴だ（ただしカボチャの種やブルーベリーなど単一の食品ばかりを食べても効果はない）。

さらに，年を取ってから地中海食に切り替えても有効らしく，わずか2年ほどで認知力補助効果が表れる場合もある。スペインのバルセロナにあるナバーラ大学の予防医学科長マルティネス・ゴンサレス（Miguel A. Martínez González）は「スーパーマンに変身はできない。認知能力を維持し，わずかに強化することも可能ではあるが，食事は魔法ではない」という。しかしそうしたわずかな改善でも，日々の生活に少なからぬ意味を持ちうる。

過去10年の研究で

かつては食事を変えても記憶力は改善できないと考えられていたが，10年ほど前からこの見方に反する研究結果が報告されるようになった。例えばコロンビア大学のスカルメアス（Nikolaos Scarmeas）らはニューヨークに住むメディケア有資格者およそ2000人（70代半ばの人が中心）の食習慣と健康状態に関する平均4年分の情報を集めて解析し，すでに心血管疾患のリスク低下と関連づけられていた地中海食が，認知力低下の遅れおよびアルツハイマー病発症率の低下と関連していると2006年に報告した。だが研究チームは単に食事内容について調べただけで臨床試験のような対照群を設けなかったので，疑問が残った。脳に生じたプラス効果はただの偶然かもしれず，あるいは地中海食を常食とする米国人に共通する別の特徴（例えば教育水準や特定の生活習慣など）がもたらした結果かもしれない。

7年後，より明確な研究結果が出た。マルティネス・ゴンサレスらがスペインの約7500人を対象に行った調査研究「PREDIMED」〔Prevention with Mediterranean Diet（地中海食を用いた疾患予防）の略〕の結果だ。この研究では被験者を無作為に2つの実験群に分け，第1群には地中海食とともに一盛りのミックスナッツを，第2群にはエキストラバージン・オリーブオイルを特にふんだんに使った地中海食を食べる生活を続けてもらった（被験者の費用負担なしにナッツとオリーブオイルを上乗せすることで，ある種の脂肪の摂取量が十分に高まって，その効果が表れるだろうと研究チームは考えた）。これに対し対照群には一般的な減量法を指示したうえで，地中海食と同様の野菜と肉，高脂肪乳製品を食べるのだが，調理にはオリーブオイルを使わず，ナッツ類も食べないよう指示した。

予想通り，地中海食を食べた2つの実験群は心血管関連の状態がかなりよくなった。そして一部被験者を抽出して認知力テストをしたところ，地中海食の2つの実験群のメンバーは減量指示だけの対照群のメンバーよりも成績がよかった。「これには驚いた」とマルティネス・ゴンサレスはいう。

興味深い発見ではあるが，やはり決定的ではない。調査開始時点で認知力のデータを取っていなかったからだ。このため，2つの実験群と対照群の間に食事以外に何か異なるものがあって，それが成績の差をもたらしている可能性が残る。

この曖昧さを排除するためマルティネス・ゴンサレスらは新たな研究を行い，2015年7月に *JAMA Internal Medicine* 誌に報告した。以前のPREDIMED研究に参加した被験者から経済的に余裕のある地域に住む300人以上を選び，まず認知力を評価しておいて，4年後の再評価と比較した。調査開始時の平均年齢は67歳。マルティネス・ゴンサレスによると，この最新の結果は前回の発見と整合した。だが，やはり決定的とはいえない。抽出調査の人数が比較的少ないからだ。それでも，ランダム化比較試験によって地中海食に認知力向上効果が認められた例はこれが初めてだと彼は指摘する。

米国人の標準的な食事と生活様式はスペイン人とかなり違うが，それでも地中海食で効果を得られるだろうか？　まだ何ともいえない。今回の研究で

対照群が食べた普通の食事は多くの米国人の食事に比べると地中海食に近いので，彼らも長年にわたって健康的な食生活を続けた結果として全般に良好な健康状態にあった可能性がある。マルティネス・ゴンサレスはむしろ米国人のほうが地中海食のメリットが大きいとみる。改善の余地が大きいからだ。それでもラッシュ大学の栄養学の専門家モリス（Martha Morris）は，この問題を決着するには米国でランダム化比較試験を行うしかないという。モリスは今後，そうした試験を実施したいと考えている。

食事を含めた生活習慣を見直す

特定の食事が頭の健康に影響することを証明するのと，多くの米国人にもっと野菜と果物，魚，オリーブオイルを食べさせるのとは，まったく別の話だ。2つの大きな障害がある。費用と，しみついた習慣だ。PREDIMED研究の場合，被験者は高価なエキストラバージン・オリーブオイルを無償で提供され，調理法も教えられた。「この知識を米国の人々に移転するには，単に食材を示すだけではだめだ」とマルティネス・ゴンサレスはいう。「食材を示し，どこで買えるかを教え，すべての栄養を伝統的な地中海食と同じに維持する調理法を教える必要がある」。第一歩となるのは，赤身肉を食べる量を減らして代わりに鶏肉にすることだろうという。だがそれでも，地中海食にするにはまだ多くのステップが残る。

PREDIMED研究と完全に同じ食事でなくても，メリットはあるようだ。モリスらは2015年2月，地中海食とほぼ同じだが米国で割安に得られる改変版を推奨する研究をオンライン公開した。「MIND食（マインド食）」と呼ばれるこの食事は葉物野菜と全粒穀物を主体とし，野菜料理を1日に2品，ベリー類を週に2品，そして地中海食では魚料理をほぼ毎日食べるのに対してこれを週に1回だけとる。

マインド食の心がけを平均4.5年続けるだけで，地中海食や他の食事に比べてアルツハイマー病のリスクが低くなる傾向をモリスらは見いだした。各集団で期間中にアルツハイマー病と診断された人数に基づいて判断した結果だ。さらに好ましいことに，マインド食は平均的な米国人にとって，経済的にも食文化のうえでも実行しやすい。この研究は「米国の人々が自費で食事内容を改めて効果を出せる」ことを示していると，ボストンにあるブリガム・アンド・ウィメ

ンズ病院とハーバード大学医学部の教授で高齢者の健康を専門にしているグロッドスタイン（Francine Grodstein）は評価する。

特定の食事がなぜ脳の健康によいのか，理由はよくわかっていない。ただ，それらの食事が心血管によいことは知られており，脳への血流と酸素供給を促進することがおそらくカギなのだろう。他の要素が寄与している可能性もある。もちろん，食事の切り替えをいつまでに行う必要があるのかや，運動や睡眠，遺伝的要因など他の要素と比較して食事にどれだけの重みがあるのかといった問題も未解明だ。

近年はそうした点に関する研究も始まっている。欧州連合（EU）は2015年，食事と運動，血圧の適正管理の組み合わせが脳の健康をどれだけ促進するかを調べるランダム化試験を始めた（高血圧は脳卒中を招く主因で，脳卒中は精神機能を大きく損なう場合がある）。どの要素が最も大きな効果を生んでいるのか特定はできないだろうが，生活習慣の変更がどれだけ重要な役割を果たしうるのか理解が進むはずだ。

期待できる材料がある。2015年6月に*Lancet*誌に報告された試験的な研究は，年を取ってから食事や生活習慣を変えても認知力低下を遅らせることができることを発見した。北欧のチームによる研究で，フィンランドの1260人を2群に分け，一方には標準的な栄養・食事を助言し，他方には特別の運動プログラムに参加するとともに地中海食の改変版を常食してもらった。期間を通じて血圧などの健康指標をモニターし，必要な場合は治療した。この結果，実験群の被験者は標準的な認知力テストの成績が有意に高かった。「この介入によって認知力の低下を防止あるいは少なくとも遅延できることがはっきりわかった」と研究論文の筆頭著者となったキビペルト（Miia Kivipelto）はいう。キビペルトはカロリンスカ研究所（ストックホルム）老人医学クリニックで研究・教育部門を率いており，予想外にもこれらの変化がわずか2年以内に見られたという。そして何よりも，スーパーパワーを必要としない。

A Turn for the Worse
回転性めまい

人工内耳と遺伝子治療による新たな治療法の試験が進みつつある

D. ヌーナン（サイエンスライター）

掲載：SCIENTIFIC AMERICAN August 2015, 日経サイエンス 2016 年 2 月号

Leaping through the air with ease and spinning in place like tops, ballet dancers are visions of the human body in action at its most spectacular and controlled. Their brains, too, appear to be special, able to evade the dizziness that normally would result from rapid pirouettes. When compared with ordinary people's brains, researchers found in a study published early this year, parts of dancers' brains involved in the perception of spinning seem less sensitive, which may help them resist vertigo.

For millions of other people, it is their whole world, not themselves, that suddenly starts to whirl. Even the simplest task, like walking across the room, may become impossible when vertigo strikes, and the condition can last for months or years. Thirty-five percent of adults older than 39 in the U.S.—69 million people—experience vertigo at one time or another, often because of damage to parts of the inner ear that sense the body's position or to the nerve that transmits that information to the brain. Whereas drugs and physical therapy can help many, tens of thousands of people do not benefit from existing treatments. "Our patients with severe loss of balance have been told over and over again that there's nothing we can do for you," says Charles Della Santina, an otolaryngologist who studies inner ear disorders and directs the Johns Hopkins Vestibular NeuroEngineering Laboratory.

Vocabulary

evade のがれる
dizziness めまい，くらくらすること
pirouette ピルエット（つま先旋回）

vertigo めまい，回転性めまい

whirl ぐるぐる回る

inner ear 内耳

physical therapy 理学療法

otolaryngologist 耳鼻咽喉科医
vestibular 前庭の

回転性めまい

Steve Bach's nightmare started in November 2013. The construction manager was at home in Parsippany, N.J. "All of a sudden the room was whipping around like a 78 record," says Bach, now age 57. He was curled up on the living room floor in a fetal position when his daughter found him and called 911. He spent the next five days in the hospital. "Sitting up in bed," he recalls, "was like sitting on top of a six-foot ladder." Bach's doctors told him that his left inner ear had been inflamed by a viral infection. He underwent six months of physical therapy to train his brain and his healthy right ear to compensate for the lost function in his left. It helped, and he returned to his job in May 2014. Even so, this spring he was still having unsteady moments as he made his way around a construction site. "Whatever is in your brain that tells you when your foot is going to hit the ground to keep you upright, I don't have 100 percent of that," he says. Vertigo can also trigger severe anxiety and depression, impair short-term memory, disrupt family life and derail careers.

Such crippling difficulties are prompting physicians to test new treatments for the most severe vertigo cases, Della Santina says. He is starting a clinical trial of prosthetic implants for the inner ear. Other doctors are experimenting with gene therapy to fix inner ear damage. And the work with dancers is beginning to reveal novel aspects of brain anatomy involved with balance, parts that could be targets for future treatments.

The ears are key to keeping us upright and stable because they hold an anatomical marvel known as the peripheral vestibular system. This is a tiny arrangement, in each ear, of fluid-filled loops, bulbs and microscopic hair cells. The hairs are topped by a membrane embedded with even tinier calcium carbonate crystals. When the head moves, the crystals pull on the hairs and combine with the other bits of anatomy to relay information about motion, direction and speed to the vestibular

Vocabulary

whip around 急に向きを変える
curl up 身を丸めて倒れる
fetal 胎児の
911 911番（北米の緊急通報用電話番号。日本の119番に相当）

inflamed 炎症を起こして
infection 感染,感染症

unsteady 不安定な,ふらふらする

short-term memory 短期記憶
derail 狂わせる,乱す

crippling ダメージの大きな,壊滅的な

clinical trial 臨床試験
prosthetic implant 埋め込み式の装具
gene therapy 遺伝子治療

peripheral vestibular system 末梢前庭系

hair cell 有毛細胞
calcium carbonate 炭酸カルシウム

vestibular nerve 前庭神経

nerve. The nerve passes it on to a region at the stem of the brain called the cerebellum, as well as other neural areas. The brain then activates various muscles and the visual system to maintain balance.

The list of things that can go wrong with this delicate system is long. Causes of inner ear vertigo include tumors, bacterial and viral infections, damage from certain antibiotics, and Meniere's disease, a chronic condition characterized by recurring bouts of vertigo, hearing loss and tinnitus that experts estimate to affect an additional five million people. The most common vestibular disorder is benign paroxysmal positional vertigo, or BPPV. It occurs when renegade crystals get loose, float into the vestibular loops and generate a false sensation of movement. Fortunately, this type of problem is usually treated effectively with physical therapy involving a repeated set of slow head movements that float the crystals out of the loops.

But physical therapy does not help everyone or, as in Bach's case, does not heal the person completely. Some patients have lost vestibular function in both ears. For them, Della Santina and his colleagues at Johns Hopkins have been developing an implant that substitutes mechanical components for damaged inner ear anatomy. Once the researchers get the green light from the U.S. Food and Drug Administration, they will begin testing their invention, called a multichannel vestibular implant, in humans. The device is modeled on the cochlear implants that have restored hearing for thousands of people since the first one was used in 1982. These implants use a microphone to pick up sound vibrations and transmit them to the brain via the auditory nerve. Instead of a microphone, a vestibular implant has two miniature motion sensors that track the movement of the head. One, a gyroscope, measures the motion of the head as a person looks up, down and around a room. The other, a linear accelerometer, measures directional movement, such as walking straight ahead or down a

Vocabulary

cerebellum 小脳

antibiotic 抗生物質
Meniere's disease メニエール病
bout 発作, 発症
tinnitus 耳鳴り
benign paroxysmal positional vertigo 良性発作性頭位めまい症
renegade 裏切りの

substitute 代替する

U.S. Food and Drug Administration 米食品医薬品局

cochlear implant 人工内耳

auditory nerve 聴覚神経

gyroscope ジャイロスコープ

linear accelerometer 線形加速度計

flight of stairs. And instead of breaking sound into different frequency components and sending them to the auditory nerve, the motion sensors send the signals connoting head position and movement to the vestibular nerve.

Results from the trial of a different vestibular implant in four patients with Meniere's disease at the University of Washington were mixed. Although it worked well initially, the effect petered out after a few months. But the Johns Hopkins device has a different design and will be used in patients with disorders other than Meniere's, so the physicians hope the outcomes will be better.

EAR GENES

Another strategy being tested in humans involves a gene that controls hair cell growth in the inner ear. During embryonic development, the *ATOH1* gene directs the creation of these cells, which are crucial for hearing and balance. The gene stops working at birth, leaving humans with a fixed number of hairs—and problems if the hairs are damaged. In an early FDA-approved clinical trial targeting balance and hearing, researchers led by Hinrich Staecker, an otolaryngologist at the University of Kansas, are injecting the gene into the ears of 45 patients with severe hearing loss, under general anesthesia. In experiments on mice with severe inner ear damage, the compound restored hair cell levels to 50 percent of normal, with some improvement in hearing. If the experimental compound, called CGF166, has similar effects in people, it could launch a new era in the treatment of vestibular disorders.

Gene therapy needs to be handled carefully; it can trigger serious immune system reactions, and patients in other experiments have died. Safety factors in this trial include a gene that can be turned on only in the targeted cells, Staecker says, and a minuscule dose that does not circulate through the body. In addition, he explains, the viral jacket around the gene, which helps it penetrate cells, has been deployed "without safety problems" in about 1,500 people in previous experiments with different genes.

Even if such research succeeds, major gaps in our basic knowledge about disabling dizziness remain. For example, doctors do not know why the ear crystals get loose in the first place. These gaps are why some researchers turned to ballet dancers. The idea is to study especially robust vestibular systems to better understand the mysteries of unhealthy ones.

Vocabulary

robust 頑健な

A team at Imperial College London used a battery of tests and brain imaging to investigate the ability of expert ballet dancers to resist vertigo while performing multiple pirouettes. The scientists studied 29 female dancers with an average of 16 years of training—the dancers started at or before age six—and compared them with female rowers. The more experienced and highly trained dancers had a lower density of neurons in parts of the cerebellum where dizziness is perceived, the group reported this year in the journal *Cerebral Cortex*. The anatomy is smaller, the researchers think, because the dancers continually suppress the perception of dizziness. During pirouettes, dancers focus their eyes on a fixed point for as long as possible. The technique, called spotting, limits the sensory signals sent to the brain. This "active effort to resist dizziness" during years of training also left the dancers in the study with a smaller, slower network of neuron connections in a part of the right hemisphere of the brain where those signals are processed.

a battery of 一連の

This kind of suppression might someday offer relief to patients with chronic vertigo, if ways can be found to develop it in nondancers using physical therapy, the scientists suggest. For thousands of patients, it would be a turn for the better.

chronic 慢性の

やすやすと宙を舞い，こまのようにまっすぐ立って回転し，バレエダンサーは人体がなしうる最も素晴らしく統制の取れた動きを見せてくれる。彼らは脳も特別なようで，通常なら急速なピルエット（つま先旋回）で目が回るのに，これを避けられる。2015年前半に発表されたある研究は，ダンサーは回転の知覚に関与する脳領域の感度が常人に比べて低いことを示した。めまいに強いのはこれが一因らしい。

一方で何百万人もの人たちが，自分の身体ではなく周囲の世界が急にぐるぐる回り始める「回転性めまい」を体験している。これに襲われると，部屋のなかを歩くといったごく簡単なこともできなくなり，その症状が数カ月あるいは数年続く場合もある。39歳超の米国人の35%（6900万人）はめまいの経験があり，多くは身体の姿勢を感知している内耳や，その情報を脳に伝える神経の損傷が原因だ。たいていは薬や理学療法でよくなるものの，既存の治療法が効かない人も数万人いる。「そうした重度の患者はどの医師にもさじを投げられてきた」と，内耳疾患を研究している耳鼻咽喉科医でジョンズ・ホプキンス大学の前庭神経工学研究室を率いているデラ・サンティーナ（Charles Della Santina）はいう。

ニュージャージー州パーシッパニーに住む建設現場監督バック（Steve Bach, 現在57歳）の悪夢は2013年11月に始まった。自宅にいたところ，「突然，部屋が78回転のSPレコードのようにぐるぐる回り始めた」という。居間の床に身を丸めて横たわっているところを娘が発見して救急車を呼んだ。彼は続く5日間を病院で過ごした。「ベッドで上体を起こすと，高さ2mのはしごの上に立っているような気がした」と振り返る。主治医によると，左の内耳がウイルス感染によって炎症を起こしているという。左内耳の失われた機能を補うため，彼は6カ月間の理学療法を受けて脳と健全な右耳を訓練した。この効果があって2014年5月に職場に復帰したが，翌年春の段階でも建設現場を巡回中にふらつくことがまだある。「脳には足がいつ地面に着くかを本人に知らせて姿勢を真っ直ぐ保っている領域があるのだろうが，私の脳ではそれが完全ではない」という。めまいはこのほか，重い不安障害やうつ病のきっかけとなり，短期記憶を損ない，家庭生活を害し，就業を困難にする恐れがある。

このため重度のめまいを治療する新しい方法が試されているとデラ・サンティーナはいう。彼は埋め込み式の人工内耳の臨床試験を始めた。内耳の障害を正す遺伝子治療を実験している医師もいる。さらに，バレエダンサーを調べた研究から平衡感覚に関与している脳の新たな側面が明らかになり，将来の治療標的になる可能性が出てきた。

　安定した姿勢を保つのに耳が欠かせないのは，「末梢前庭系」という驚異の構造が内耳に存在するからだ。この小さな構造は左右それぞれの耳にあり，体液で満たされた三半規管や小さな有毛細胞などからなる。有毛細胞の感覚毛は，さらに小さな炭酸カルシウムの耳石が乗った膜に埋め込まれている。頭が動くと耳石の動きにつれて感覚毛が曲がり，その他の構造の働きもあって，動きとその方向および速度に関する情報が前庭神経に伝えられる。前庭神経はこれを小脳などいくつかの領域に伝え，脳はこれに基づいて様々な筋肉と視覚系を働かせて身体のバランスを維持する。

　この微妙なシステムが崩れると様々な不調を来す。内耳性めまいの原因は腫瘍，細菌やウイルスによる感染症，ある種の抗生物質がもたらすダメージなど。またこのほかに，繰り返し生じるめまいと聴力低下，耳鳴りを特徴とする慢性疾患「メニエール病」の患者が500万人に達すると推定されている。最も一般的に見られる前庭系疾患は「良性発作性頭位めまい症（BPPV）」で，耳石が本来の膜をはずれて半規管に入り込み，これが液中を漂って，実際には身体が動いていないのに動きの感覚を生み出す。幸いこのタイプの障害は，頭を何度もゆっくり動かして耳石を半規管から外に出す理学療法によって，効果的に治療できる場合が多い。

　だが理学療法が全員に有効というわけではなく，バックのように完治しない例もある。また，両耳の前庭機能を失った患者もいる。そうした患者のために，デラ・サンティーナらジョンズ・ホプキンズ大学のチームは内耳の機構部品を代替する埋め込み型装置を開発してきた。米食品医薬品局（FDA）の許可が出たら，「多チャンネル前庭インプラント」というこの装置を患者で試す計画だ。この機器は，1982年に実用化してから多くの人の聴力を回復してきた人工内耳をもとに設計された。人工内耳はマイクロホンで拾った音の振動を聴覚神経を通じ

て脳に送る。これに対し前庭インプラントは，マイクではなく動きを検知する2種類の小型センサーを用いて頭部の動きを追跡する。1つはジャイロスコープで，部屋のなかを見回す際の頭部の上下左右への動きを計測する。もう1つは線形加速度計で，真っ直ぐ前に向かって歩いているのか階段を下りているのかなど，動きの向きを検出する。そして，人工内耳が音を振動数別に分けて聴覚神経に伝えるのに対し，これらの運動センサーは頭部の姿勢と動きを表す信号を前庭神経に送り出す。

別タイプの前庭インプラントを4人のメニエール病患者で試したワシントン大学（シアトル）による臨床試験の成績はいまひとつだった。当初はうまく働いたものの，数カ月で効果が薄れたのだ。だがジョンズ・ホプキンズ大学の装置は設計が異なり，メニエール病以外の患者も対象になるので，よりよい結果が出るだろうと研究チームは期待している。

有毛細胞を作る遺伝子治療

一方の遺伝子治療は内耳の有毛細胞の成長を制御する遺伝子に注目したものだ。聴覚と平衡感覚に不可欠なこの細胞は胎児の段階で作られ，*ATOH1*という遺伝子がその生成を指令している。出生とともにこの遺伝子の働きは止まり，そこで有毛細胞の数は決まる。つまり，後にそれらが損傷しても補充はない。カンザス大学の耳鼻咽喉科医シュテッカー（Hinrich Staecker）が率いるチームは重度の難聴患者45人を対象に，この遺伝子を全身麻酔下で内耳に注射する臨床試験を行っている。これに先立つ動物実験では，内耳にひどい傷害を負ったマウスの有毛細胞が正常値の50%に回復し，聴力も改善した。「CGF166」と呼ばれるこの薬剤が人間でも同様の効果を発揮すれば，前庭疾患の治療に新時代が開けるだろう。

ただし遺伝子治療は慎重に行う必要がある。重大な免疫応答を引き起こす場合があり，初期の別の臨床試験では患者が死亡した例もある。今回の試験では，目的の細胞だけで遺伝子が発現するよう調整してあるほか，投与量がわずかなので全身に回ることはないとシュテッカーはいう。また，遺伝子を細胞に導入するベクターとして使っているウイルスは，これまでの別の臨床試験で約1500人に投与されて「安全上何の問題もなかった」ものだという。

これらの研究が成功しても、回転性めまいに関する基本的知識には大きな空白が残る。例えば耳石が所定の位置からなぜはずれてしまうのか、そもそもわかっていない。一部の研究者がバレエダンサーに目を向けたのは、こうした基本的な謎を解明するためだ。飛び抜けて頑健な前庭系を調べることで、その謎に迫る。

ロンドン大学インペリアルカレッジのチームは一連のテストと脳画像撮影によって、熟練のダンサーが何回ピルエットを演じても目が回らないのはなぜかを調べている。平均で16年の経験を持つ29人のバレリーナ（6歳前にバレエを始めた人たち）を調べ、女性ボート選手と比較した。この結果、経験が長く訓練を積んだダンサーほど、めまいを感知する小脳の領域におけるニューロンの密度が低いことを突き止め、2015年の *Cerebral Cortex* 誌に報告した。めまいの感知を常に抑制してきたことによって、この脳領域が小さくなったのだろうと研究チームはみている。ピルエットの間、ダンサーはできる限り一点を見つめるよう努める。「スポッティング」というこのテクニックによって、脳に送られる感覚シグナルが抑制される。この「めまいに積極的に対抗する活動」を長年の訓練を通じて繰り返したことによって、被験者のバレリーナたちは、これらの信号が処理される右脳の一部領域における神経回路が小規模で、処理速度は遅くなっていた。

こうした抑制を理学療法によってダンサー以外の人に引き起こす方法が見つかれば、慢性的なめまいの緩和につながるかもしれない。めまいに苦しむ大勢の患者にとって、それは世界がよい方向へ回ることを意味するだろう。

Can We Stop Aging?
老化を止められるか？

一部の研究者はそれが可能になるとみている。少なくともしばらくの間は

K. ワイントラウブ（サイエンスライター）

掲載：SCIENTIFIC AMERICAN July 2015, 日経サイエンス 2016 年 1 月号

The majority of older Americans live out their final years with at least one or two chronic ailments, such as arthritis, diabetes, heart disease or stroke. The longer their body clock ticks, the more disabling conditions they face. Doctors and drug companies traditionally treat each of these aging-related diseases as it arises. But a small group of scientists have begun championing a bold new approach. They think it is possible to stop or even rewind the body's internal chronometer so that all these diseases will arrive later or not at all.

Studies of centenarians suggest the feat is achievable. Most of these individuals live that long because they have somehow avoided most of the diseases that burden other folks in their 70s and 80s, says Nir Barzilai, director of the Institute for Aging Research at the Albert Einstein College of Medicine. Nor does a centenarian's unusual longevity result in an end-of-life decline that lasts longer than anyone else's. In fact, Barzilai notes, research on hundreds of "super agers" suggests exactly the opposite. For them, illness typically starts later and arrives closer to the end. "They live, live, live and then die one day," he says.

Researchers have already developed various techniques to increase the life span of yeast, worms, flies, rats and perhaps monkeys. Adapting these measures to people

Vocabulary

chronic ailment 慢性疾患
arthritis 関節炎
diabetes 糖尿病
stroke 脳卒中

champion 〜の闘士として働く，〜に挑戦する

chronometer クロノメーター（高精度の時計）

centenarian 100 歳以上の人

longevity 長寿

yeast 酵母

seems like the next logical step. "There's an emerging consensus that it's time to take what we've learned from aging [research] and begin to translate that into helping humans," says Brian Kennedy, CEO and president of the Buck Institute for Research on Aging, an independent research group in Novato, Calif.

Delaying the aging process by even a few years could offer enormous social benefits as populations around the globe grow increasingly older. The U.S. Census Bureau estimates that one in five Americans will be older than 65 by 2030—up from one in seven in 2014. In 2013 an estimated 44 million people around the world suffered from dementia. That number is expected to jump to nearly 76 million in 2030 and 135 million in 2050—with not nearly enough younger people in a position to be able to take care of them.

Among the handful of approaches that researchers are studying, three stand out. Still unclear: whether the potential benefits outweigh the risks of the treatments.

EVIDENCE

Of course, to conclusively determine whether a treatment works, investigators need a definition of aging and a way to measure the process. They have neither. If a kidney cell divided yesterday, is it one day old or as old as the person in whom it resides? Still, research over the past decade has offered several hints that the damaging aspects of aging—however you define it—can be slowed.

In a 2005 study, Thomas Rando, director of the Paul F. Glenn Center for the Biology of Aging at Stanford University, showed that an elderly mouse whose bloodstream was surgically linked to a young mouse recovered its youthful wound-healing powers. Somehow the older rodent's stem cells, which are responsible for replacing damaged cells, became more effective at giving rise to new

Vocabulary

Census Bureau 国勢調査局

dementia 認知症

stand out 突出する, 目立つ
outweigh しのぐ, 上回る

surgically 外科的に

stem cell 幹細胞

tissue. Harvard University biologist Amy Wagers has since found a protein, dubbed GDF11, in the blood that may have contributed to the faster healing. Her experiments, published in *Science* in 2014, found more of the protein in younger mice than in older ones; when injected in older mice, GDF11 appeared to restore muscles to their youthful structure and strength. A new study, in *Cell Metabolism*, calls that finding into question, however, suggesting that GDF11 increases with age (and may even inhibit muscle restoration) and that some other factor must make the cells act younger.

A second approach consists of examining about 20 currently existing medications and nutritional supplements at a level of detail that has never before been possible to see whether they might actually affect the aging process. For example, researchers at Cardiff University in Wales and their colleagues reported in 2014 that patients with type 2 diabetes who took the drug metformin lived, on average, 15 percent longer than a group of healthy people who did not suffer from the metabolic disorder but were similar in nearly all other respects. Scientists speculate that metformin interferes with a normal aging process, called glycation, in which glucose combines with proteins and other important molecules, gumming up their normal workings. The metformin finding is particularly striking because people who have diabetes, even if it is well controlled, typically have somewhat shorter life spans than their healthy counterparts.

Meanwhile, in a study of 218 adults published late last year in *Science Translational Medicine*, researchers at pharmaceutical company Novartis showed that a compound called everolimus, which is chemically similar to rapamycin (a drug used to prevent kidney rejection in transplants), improved the effectiveness of the flu shot in people older than 65.

Vocabulary

call ~ into question ～に疑問を投げかける

type 2 diabetes 2型糖尿病
metformin メトホルミン(糖尿病治療薬の一種)
metabolic disorder 代謝性疾患

glycation 糖化, 糖化反応
glucose グルコース, ブドウ糖

gum up 狂わせる

everolimus エベロリムス(免疫抑制剤の一種)
rapamycin ラパマイシン
flu shot インフルエンザ予防ワクチンの接種

As individuals age, their immune systems do not mount as strong an antibody response to the inactivated virus in the vaccine as they once did; thus, older people are more likely to get sick if they later encounter a real flu virus. Tests showed that study patients given everolimus had a higher concentration of germ-fighting antibodies in their blood than their untreated counterparts. Investigators interpreted this finding as a sign that the drug had rejuvenated the subjects' immune systems.

As with any drug, side effects were an issue. Members of the treated group were more likely to develop ulcers in their mouth, which may limit the widespread usefulness of the medication for treating aging. Cost may be another factor; everolimus, which was approved by the U.S. Food and Drug Administration for its cancer-fighting properties, costs more than $7,000 a month at doses appropriate for cancer. Not yet known: how much everolimus would cost and how long it would be needed, if used as an antiaging drug.

Nevertheless, the results support the idea that aging can be slowed. Indeed, everolimus and other rapamycinlike drugs have been shown to dramatically extend the life span of mice, preventing diseases such as cancer and reversing age-related changes to the blood, liver, metabolism and immune system.

A third, completely different approach involves diet. Restricting the consumption of calories was long ago shown to help mice to live longer. Whether limiting food intake (without causing malnutrition) might benefit humans as well is not so clear. For one thing, very few people can or want to maintain such low-calorie diets for the decades needed to prove definitively that this approach works. But it may turn out that such drastic steps are unnecessary. Valter Longo, director of the Longevity Institute at the University of Southern California, has shown that he can extend the life span of mice merely by limiting their food on alternate days or by cutting down on the amount of protein they consume. Such intermittent fasting may turn out to be more palatable for people, although its benefits remain unproved.

CAVEATS

Living longer may come with trade-offs. Making old cells young again will mean they will start dividing again. Controlled cell division equals youthfulness; uncontrolled cell division equals cancer. But at the moment, scientists are not sure if they can do one without the other.

Figuring out the right timing for treatment is also complicated. If the goal is to prevent multiple diseases of aging, do you start your antiaging therapies when the first disease hits? The second? "Once you're broken, it's really hard to put you back together. It's going to be easier to keep people healthy," Kennedy says. So it probably makes more sense to start treatment years earlier, during a healthy middle age. But the research needed to prove that supposition would take decades.

If various diseases can be pushed off, the next obvious question is by how long. James Kirkland, who directs the Mayo Clinic's Robert and Arlene Kogod Center on Aging in Rochester, Minn., says it will take at least another 20 years of study to answer that question. Scientists have successfully extended the life span of worms eightfold and added a year of life to three-year-old lab mice. Would these advances translate into an 80-year-old person living five or six centuries or even an extra 30 years? Or would they get just one more year? Life extension in people is likely to be more modest than in yeast, worms, flies or mice, Rando says. Previous research has suggested that lower-order creatures benefit the most from longevity efforts—with yeast, for instance, deriving a greater benefit in caloric-restriction experiments than mammals. "The closer you get to humans, the smaller the effect" on life span, he says. And what magnitude of benefit would someone need to justify taking—and paying for—such a treatment? "Do you take a drug your whole life hoping to live 4 percent longer or 7 percent longer?" Rando asks.

What, if anything, do antiaging investigators themselves do to try to slow their own aging? The half a dozen scientists interviewed for this article all said that they make concerted efforts to extend their own life span. One was grateful for a diagnosis of prediabetes, which meant a legitimate prescription for metformin. The research is getting so solid, Kennedy says, that he is having a tougher time convincing himself not to take some drugs than to take them.

All the experts say they try to live healthy lives, aside from enduring high-pressured jobs. They try to get close to eight hours of sleep, eat moderate amounts of nutritious foods and get lots of exercise. None of them smokes. Most Americans, unfortunately, do not follow such healthy habits. The greatest irony would be to discover that a pill is not, in the end, any more effective than the healthy habits we already ignore.

Vocabulary

concerted 申し合わせた, 一致した
prediabetes 前糖尿病

米国の高齢者の多くは，関節炎や糖尿病，心臓病，脳卒中の後遺症など，慢性疾患を抱えて晩年を送る。年を取るほど，症状は重くなる。医師はこれまで，これらの加齢関連疾患を発症後に治療してきた。これに対し一部の科学者は大胆な新アプローチに挑み始めている。人体の時計を止め，さらには逆転させることで，こうした疾患の発症を遅らせ，あるいは回避できるという考え方だ。

百歳以上の超長寿者を調べた研究は，この離れ業が可能であることを示唆している。超長寿者の多くは通常なら70代や80代でかかる疾患をなぜか回避して長生きしていると，アルバート・アインシュタイン医科大学・加齢研究所の所長バージライ（Nir Barzilai）はいう。また，超長寿者は人生終盤の衰退期が長く続くのではない。正反対を多くの研究が示している。病気の発症が遅く，天寿に近い。「生きて生きて生き続けて，ある日に亡くなる。ピンピンコロリなのだ」。

酵母や線虫，ハエ，ネズミ，サルについては，寿命を延ばす様々な方法がすでに開発されている。次はそれらを人間に適用することだろう。「加齢研究を通じて得られた知見を人間に役立てる時がきたとの認識が広がっている」と非営利の民間研究機関であるバック加齢研究所（カリフォルニア州ノバト）の代表ケネディ（Brian Kennedy）はいう。

世界で高齢化が進むなか，加齢を2〜3年でも遅らせることができれば社会的利益は非常に大きい。米国勢調査局の推定によると，2014年に米国人の7人に1人が65歳超の高齢者だったものが，2030年には5人に1人となる。2013年時点で世界の推定4400万人が認知症を患っているが，これが2030年には7600万人に近づき，2050年には1億3500万人に跳ね上がると予想される。

現在研究されているアプローチのうち，特に3つが注目されている。不透明なのは，それらから得られるメリットがリスクを上回るかどうかだ。

確かな証拠を求めて

治療の有効性を正確に評価するには，当然ながら加齢の定義と計測法が必要だ。だが，どちらも存在しない。腎臓の細胞が昨日に細胞分裂した場合，その細胞の年齢は1日なのか，その細胞の主である人と同じ年齢なのか？　だが，

加齢の正確な定義とは無関係に，有害な影響が生じるのを遅延できる手がかりが過去10年の研究でいくつか得られた。

2005年，スタンフォード大学・加齢生物学センター所長のランド（Thomas Rando）は，高齢のマウスと若いマウスの血管を外科手術でつなぐと高齢マウスの創傷治癒能力が回復することを示した。古い細胞や損傷した細胞を置き換えて修復する幹細胞の働きがなぜか強まり，新組織を効果的に作り出せるようになった。ハーバード大学の生物学者ウェイジャーズ（Amy Wagers）はその後，この治癒能向上をもたらしているとみられるタンパク質GDF11を血液中に発見した。彼女が2014年に*Science*誌に発表した研究によると，このタンパク質は若いマウスに多く，高齢マウスにGDF11を注射で与えると筋肉の組織構造と強度が回復した。ただ，その後*Cell Metabolism*誌に掲載された別の研究はこの発見に疑問を投げかけ，GDF11は加齢とともにむしろ増え（そして筋肉の再生を阻害するらしく），幹細胞の働きを強めているのは何か別の因子に違いないとみている。

2番目のアプローチは，既存の医薬品と栄養補助剤およそ20種を詳しく調べ，これらが加齢プロセスに本当に影響を与えているかどうかを見極める。例えば英カーディフ大学の研究チームは2014年，「メトホルミン」という薬を服用している2型糖尿病患者が，健康な対照群（糖尿病以外は実験群とほぼ同じ特性の人たち）に比べ平均で15％長生きであると報告した。「糖化反応」という正常な加齢プロセスをメトホルミンが阻害するのだと考えられている。糖化反応はブドウ糖がタンパク質などの重要な分子に結びついて，本来の働きを損なう反応だ。一般に糖尿病患者は，血糖値をうまくコントロールしていても健康な人よりもいくぶん短命であるとされてきたので，この発見は実に衝撃的といえる。

一方，製薬企業ノバルティスは218人の成人に「エベロリムス」を投与した研究を2014年の*Science Translational Medicine*誌に報告した。エベロリムスはラパマイシン（腎臓移植の拒絶反応抑制に使われる薬）と化学的に似た薬で，これを投与された65歳以上の人はインフルエンザ予防ワクチンの効果が高まった。

年を取ると一般に免疫系の反応が鈍り，ワクチン中の不活化ウイルスに対する抗体ができにくくなる。このため接種後に本当のウイルスに遭遇すると

発病しやすい。だが，エベロリムスを投与された人たちは対照群に比べ，ウイルスに対する抗体の血中濃度が高くなった。エベロリムスが被験者の免疫系を若返らせたしるしだと，研究チームは解釈している。

他の薬と同様，副作用が問題だ。エベロリムスを投与された人は口に潰瘍ができやすく，アンチエイジング目的で幅広く使うわけにはいかないかもしれない。費用も問題になりそうだ。この薬は抗がん剤として米食品医薬品局(FDA)の認可を得ており，抗がん剤として投与される場合の費用は月に7000ドルを超える。もっとも，抗老化薬として使う場合の投与量と期間，費用はまだ不明だ。

ではあるが，この結果は加齢を遅らせうるという考えを支持している。実際，マウスではエベロリムスや他のラパマイシンに似た薬が寿命を劇的に延ばすことが示されている。がんなどの疾患を防ぐほか，血液や肝臓，代謝，免疫系に生じる加齢関連の変化を逆転させる。

3番目のアプローチは食事に注目する。カロリー摂取を制限したマウスが長生きすることが示されて久しい。栄養不良にならない程度に食事量を減らすことが人間にも同様の利益をもたらすかどうかは不明確だ。有効性を確証するには低カロリー食を何十年も続けて影響を見る必要があるが，被験者になろうという人がほとんどいないのだ。だが，そんな厳しい食事制限は不要と判明するかもしれない。南カリフォルニア大学長寿研究所の所長ロンゴ(Valter Longo)はマウスの実験で，1日おきに食事制限するかタンパク質の摂取量を抑えるだけで寿命を延ばせることを示した。こうした間欠的な絶食のほうが人々には好ましいだろう。ただ，効果のほどはまだ証明されていない。

数々の問題点

長生きは代償を伴う可能性がある。古い細胞を若返らせるということは，再び細胞分裂が始まるということ。制御された細胞分裂は若さの証拠だが，とめどない増殖はがんと同じだ。前者だけを実現できるのか，現時点では不明だ。

治療を始める適切なタイミングも難しい。加齢に伴う複数の病気を防ぐのが目的である場合，アンチエイジング療法を始めるのは最初の疾患が発症し

たときがよいのか，それとも 2 つ目が発症した時点か？ 「ひとたび倒れたら，完全に回復するのは実に難しい。健康をずっと維持するほうが容易だろう」とケネディはいう。だから治療開始をもっと早く，中年の健康なころから始めるのがおそらく妥当だ。しかし，この仮説を証明する研究には数十年かかるだろう。

様々な病気の発症を遅延できるとして，次の問題はどこまで寿命を延ばせるかだ。メイヨークリニックのロバート・アンド・アイリーン・コゴッド加齢研究所の所長カークランド（James Kirkland）は，これに答えを出すには少なくともあと 20 年はかかるという。科学者たちはすでに線虫の寿命を 8 倍に延ばし，平均寿命 3 年の実験用マウスをもう 1 年長生きさせることに成功している。これは，80 歳の老人が 600 年生きられるとか，余命を 30 年延ばせると読み替えられるのか？ あるいはあと 1 年だけ生きられるのか？ 人間で可能になる寿命延長は酵母や線虫，ハエ，マウスの場合よりもささやかだろうとランドはいう。下等な生物ほど大きな効果が生じることが示唆されている。例えば酵母はカロリー制限実験で哺乳動物よりも寿命を大きく延ばした。「人間に近づくほど，効果は小さくなる」。さらに，どれだけ寿命を延ばせたら，こうした治療を受ける（そして費用を負担する）価値があるのか。「寿命が 4％延びるなら薬をのむのか，7％延びないと納得できないだろうか？」とランドは問う。

それはともかく，アンチエイジング研究者自身は老化を抑えるために何を実行しているのだろう？ 今回私が取材した数人の科学者はみな，長生きのために計画的な努力をしていると語った。なかには前糖尿病状態と診断され，メトホルミンを正当に処方されることができてむしろ幸いだという人もいた。アンチエイジング薬研究の信頼性は高まっており，そうした薬の服用を控えるよう自制するのが以前よりも難しくなっているとケネディはいう。

私が取材した専門家はみな，プレッシャーの大きな仕事に耐えることを別として，健康的な生活を心がけている。なるべく 8 時間近く眠り，栄養に富んだ適量の食事をし，たくさん運動する。喫煙者はいない。だが多くの米国人は残念ながら，そこまで健康的な生活習慣にはない。アンチエイジング薬の効果が，私たちがいま無視している健康的な生活習慣がもたらす効果と結局は変わらないことが判明する――そうなったら最大の皮肉といえよう。

潜むリスク

Cold Comfort
お寒い冷却療法

The Acupuncture Myth
鍼治療の神話

Overreaction
過剰反応のアレルギー検査

Deadly Drug Combinations
危うい薬の飲み合わせ

When DNA Means "Do Not Ask"
遺伝子検査のジレンマ

Cold Comfort
お寒い冷却療法

全身の過冷却で関節痛からお腹のたるみまで様々な問題を解決するというが
その根拠は薄氷のように頼りない

D. F. マロン (SCIENTIFIC AMERICAN 編集部)

掲載：SCIENTIFIC AMERICAN January 2017, 日経サイエンス 2017 年 12 月号

The day Phil Mackenzie decided to expose his almost naked body to gas colder than the lowest natural temperature ever recorded on Earth started like any other day. The professional rugby player woke up and headed to the playing field in Manchester, England, for his usual grueling workout. He ran passing and kicking drills. He was repeatedly tackled. He lifted weights. By the end of practice he was exhausted. Usually Mackenzie would head back to the locker room and soothe his sore body with a hot shower. On this day, however, an enclosed pod resembling a massive standing tanning bed beckoned from the nearby parking lot. Mackenzie and a couple of his teammates stepped inside. Frigid gas started to swirl around them.

Mackenzie had wanted to try this procedure, called whole-body cryotherapy, specifically to ease his achy joints. But he says that after receiving multiple two-minute sessions spread out over several days he saw other benefits, too. "I felt refreshed right away. My sleep was better," he recalls. Soon the treatments became routine: Mackenzie would go four times a week to chill out amid the icy vapors, wearing nothing but his spandex shorts, gloves, socks, slippers and headband to protect against frostbite. Most of his teammates also adopted the regimen.

Vocabulary

grueling 厳しい, へとへとにさせる
workout 練習

head back 帰る, 戻る
soothe なだめる, いやす
sore 痛む
pod さや, 容器
tanning bed 日焼けベッド
beckon 手招きする, 待ち受ける
swirl 渦巻く

whole-body cryotherapy 全身クライオセラピー, 全身冷却療法
achy 痛みのある

frostbite 凍傷

In fact, there was usually a line for the pod after practice.

Mackenzie and his fellow rugby players are hardly the only devotees of cryotherapy. Star athletes, including Kobe Bryant and LeBron James, have turned to it. Reportedly, Hollywood A-listers such as Daniel Craig and Jennifer Aniston have, too. The market for these devices is beginning to burgeon in the U.S., with sports teams snapping them up to condition their players and spas and wellness centers installing them for clients looking to relax, lose weight and fight signs of aging. One large U.S. distributor of whole-body cryotherapy machines, Dallas-based CryoUSA, says it has installed more than 200 units across the country since 2011, half of them in 2015. The company expects that the 2016 tally will show an even sharper uptick in sales.

Yet the science behind these devices is decidedly lackluster. In July the U.S. Food and Drug Administration issued a warning stating that there is no evidence these technologies help to ease muscle aches, insomnia or anxiety or provide any other medical benefit. Instead, it said, they may cause frostbite, burns, eye damage or even asphyxiation. In a statement to *Scientific American* the agency added, "The FDA has not approved or cleared any whole-body cryotherapy devices, and we do not have the necessary evidence to substantiate any medical claims being made for these devices." The agency based its warning on its own informal review of published literature and generally recognized hazards associated with exposure to the gas that creates the cold conditions in the treatment chamber. Adding insult to injury, cryotherapy is pricey. A package of five two-minute sessions can cost several hundred dollars.

A CHILL IN THE AIR

The notion of supercooling the entire body for therapeutic reasons got its start in Japan during the late

Vocabulary

devotee 熱中している人
turn to ～に頼る

A-lister 有名人, セレブ

burgeon 急成長する
snap up 先を争って買う

tally 計数結果, 総量
uptick 増加, 上昇

lackluster 精彩のない
U.S. Food and Drug Administration 米食品医薬品局

insomnia 不眠症

asphyxiation 窒息

substantiate 実証する

add insult to injury ひどいめにあわせたうえに侮辱を加える

therapeutic 治療上の

1970s, when it was touted as a potential way to relieve joint pain in patients with multiple sclerosis or rheumatoid arthritis. It then gained traction in western Europe in the 1990s. Only recently, in the past decade, has it risen to prominence in the U.S. and Australia. As the practice has spread, the list of ailments that it can supposedly address has exploded. According to the latest marketing claims, it can treat not only pain but conditions ranging from asthma to Alzheimer's disease.

The logic of whole-body cryotherapy stems from the widely accepted science underlying standard-issue cold therapy, which uses ice packs and ice-water baths to treat acute soft-tissue injuries. Doctors will typically recommend icing as part of a care regimen for a sprained or strained ankle, for example. Clinical studies have found that applying ice to an injury site for some five to 15 minutes can lower skin temperature to less than 55 degrees Fahrenheit, which slows and thus dulls pain signals from affected nerves. Ice may help in another way, too. Animal studies suggest that it combats inflammation after injury by decreasing the number of white blood cells moving to the injury site, among other mechanisms, says Chris Bleakley, a sports medicine researcher at Ulster University in Northern Ireland. (Prolonged inflammation can extend pain, decrease range of motion and impair the blood flow around the damaged area.)

But whether cryotherapy can actually produce those same benefits is uncertain at best. Unlike run-of-the-mill cold therapy, it uses gasified liquid nitrogen to cool the air around recipients who stand in an enclosed chamber to temperatures below –200 degrees F. Although the gas temperature is much colder than ice, the cold from ice applied directly to the body has a better chance of penetrating through layers of skin and fat to reach the target soft tissue than does icy gas that swirls around the skin but is not pressed against it, making chilling of deeper parts of

Vocabulary

tout 大げさに推奨する
multiple sclerosis 多発性硬化症
rheumatoid arthritis 関節リウマチ
prominence 目立つこと
ailment （慢性的な）病気
address 対処する
asthma 喘息

standard-issue 標準的な
ice pack 氷嚢

soft-tissue 軟部組織
sprain くじく, ひねる
strain くじく, 曲げる

dull 鈍らせる

inflammation 炎症

white blood cell 白血球

impair 損なう

run-of-the-mill 普通の, ありふれた
nitrogen 窒素

apply 当てる

the body harder to achieve.

Vocabulary

Indeed, a 2014 analysis of preexisting ice, cold-water and whole-body cryotherapy studies, carried out by Bleakley and by other researchers, found that ice packs delivered the biggest reductions in skin temperature and intramuscular temperature: a 10-minute ice-pack application cooled skin between 32 and 47 degrees F, for example. Three minutes of whole-body cryotherapy, however—the average time manufacturers recommend to protect user safety—resulted in a lesser reduction, ranging between six and 35 degrees F.

intramuscular 筋肉内の

Because whole-body cryotherapy is not as effective at cooling intramuscular temperatures, it is unlikely to slow pain signals as effectively as ice does or to cool soft tissues enough to quell inflammation, Bleakley says.

quell 抑える, 鎮める

Other studies compound these doubts. In the gold standard approach to evaluating efficacy of a given therapy, participants are randomly designated to receive the treatment in question, a different one or none at all. To date, researchers have conducted four such randomized control trials of whole-body cryotherapy. In an exhaustive examination of those studies, exercise physiologist Joe Costello of the University of Portsmouth in England, along with Bleakley and others, found no significant benefit to the treatment. "There is insufficient evidence to prove whether whole-body cryotherapy reduces muscle soreness or improves recovery after exercise compared to … no intervention," he states.

compound 問題を大きくする, 悪化させる
efficacy 効力, 効き目
participant (実験の)参加者, 被験者
randomized control trial
　▶ Technical Terms
exhaustive 徹底的な
exercise physiologist 運動生理学者

muscle soreness 筋肉痛

Technical Terms
ランダム化比較試験(**randomized control trial**)　有効性を調べようとする治療を受ける実験群と, その処置を受けない対照群(在来の治療法を受ける, 何も処置されない, プラセボを投与されるなど)に被験者を無作為に振り分けて, 結果を比較する試験。最も客観的な評価手法として, 臨床試験の王道とされる。さらにバイアスを排除するため, 被験者がどちらに振り分けられたかを被験者本人にも試験にあたる医師にも知らせずに行う場合があり, これを二重盲検法(double blind test)という。

Those four trials, as well as Costello's assessment of them, are not the final word. They were very small, totaling just 64 subjects. And because all but four of the subjects were men, with an average age in their early 20s, it is impossible to say whether the putative panacea might affect women or older people differently.

UNANSWERED QUESTIONS

The shortcomings of these trials are emblematic of the poor state of the science of whole-body cryotherapy. Most studies of the treatment involve "very small numbers" of participants and have "methodological flaws" such as the lack of a control group, Bleakley says. "Sports scientists really need to pick up this area and align it with the quality of studies in wider medicine," he asserts.

As for the effects of whole-body cryotherapy on all the other ailments it can purportedly address beyond athletic injuries, the science is virtually nonexistent. The claims have not been subjected to the rigors of a randomized trial. Nor do researchers have definitive answers about whether exposure to gasified liquid nitrogen produces beneficial effects on heart rate, blood pressure or metabolism—effects that, if they occurred, might help ease anxiety, treat migraines or fuel weight loss, among other aims.

Mark Murdock, managing partner at CryoUSA, does not dispute that whole-body cryotherapy lacks evidence for many of the uses claimed for it. The company promotes the devices for reducing pain and inflammation and increasing energy, but in his view, that use provides "comfort," not medical assistance. He adds that medical claims, such as that the devices can drive weight loss, are "crazy." He also says he supports the FDA's decision to release the warning it issued in July and thinks the agency should ultimately step in to regulate the industry and curb such assertions.

Not only are the supposed benefits of cryotherapy chambers unproved but scientists also lack a clear understanding of any risks they might pose. No studies have focused on adverse effects. And not all whole-body cryotherapy is created equal: treatments vary in duration, temperature and which body parts are spared contact with the subzero vapors. How long a person is exposed, at what temperature and under what conditions matter for safety, says Naresh Rao, the USA Water Polo Olympic team's physician.

Nevertheless, the notion of treating what ails us with a stint inside a glorified freezer has a powerful allure. Recipients report positive effects anecdotally, but the lack of evidence to support these claims suggests they may simply stem from belief in the treatment—the placebo effect. Rao, who is also a doctor of osteopathy (a field that supplements traditional medical care with holistic treatments), says that although he would not choose cryotherapy as first-line treatment for injured athletes, he supports his patients who want to use it—even if the benefits are subjective at best. Yet, he notes, "I do think it needs to be medically regulated. I wouldn't say it's ready for a consumer coming off the street." People with heart issues or uncontrolled hypertension, for example, should not seek out cryotherapy, he warns, because sudden exposure to such cold temperatures could trigger heart attacks or other serious health complications in these individuals.

Some researchers are still hoping for good news about cryotherapy's efficacy. Rebeccah Rodriguez, a Science Board member of the President's Council on Fitness, Sports & Nutrition, an osteopath and the physician for the San Diego Breakers rugby team, is among them. She plans to start a study in 2017 focused on evaluating cryotherapy chambers for facilitating recuperation from concussions. And a research team in Marseilles is conducting a preliminary study to assess whether whole-body cryotherapy has

Vocabulary

adverse effect 有害な効果，副作用

spare 〜しないでおく

ail 悩ます，苦しめる
stint お勤め
allure 魅力
anecdotally 逸話的に

placebo effect プラセボ効果
osteopathy 整骨療法

hypertension 高血圧

heart attack 心臓発作
complication 合併症

osteopath 整骨医

recuperation 回復
concussion 脳震盪

anti-inflammatory effects that could make it a viable alternative to popping traditional nonsteroidal anti-inflammatory drugs (known as NSAIDs).

"There is much work to be done," Ulster's Bleakley says. Only large randomized controlled studies can gauge the efficacy of whole-body cryotherapy—and arm consumers with the cold, hard facts.

Vocabulary

nonsteroidal anti-inflammatory drug 非ステロイド性抗炎症薬

arm 武装させる
cold 客観的な, 冷徹な

地球上で自然に生じる最低気温よりも冷たいガスにほとんど裸の身をさらそうとプロのラグビー選手マッケンジー（Phil Mackenzie）が決めたその日は，いつも通りに始まった。朝起き，英マンチェスターにあるグラウンドでの厳しい練習に向かう。パスとキックを繰り返し，何度もタックルを受けた。さらにウエートトレーニング。練習が終わるころにはへとへとだった。通常ならロッカールームで熱いシャワーを浴びて痛んだ肉体をいやすのだが，この日は近くの駐車場に大きな日焼けベッドを縦にしたような装置が運び込まれている。マッケンジーと数人のチームメートが順にその内部に入ると，身の周りで冷たいガスが渦巻いた。

マッケンジーはこの「全身クライオセラピー（全身冷却療法）」をかねて試したいと望んでいた。特に関節痛を和らげたかったのだが，1回2分間の処置を数日間で何度も受けたところ，ほかにも効果があったという。「リフレッシュした感じ。よく眠れるようになった」。この処置はやがて定番になった。彼は週に4回，凍傷防止用のショーツと手袋，ソックス，スリッパ，ヘッドバンドだけを着用して冷気に身をさらしている。チームメートの多くもこの療法を採用した。練習後には装置の前に順番待ちの列ができる。

冷却療法のファンは彼らだけではない。コービー・ブライアントやレブロン・ジェームズらバスケットボールのスター選手も使っている。伝えられるところでは，ダニエル・クレイグやジェニファー・アニストンといった映画スターも。米国における冷却装置の市場は急拡大を始めた。プロスポーツチームが選手の調

整用に先を争って購入し，スパやフィットネスセンターはリラックスと減量，若返り効果を求める顧客向けに導入している。装置販売大手のクライオ USA（本社ダラス）は 2011 年以降で全米に 200 台以上を納入したという。その半数は 2015 年の実績で，2016 年はさらに大きな伸びになると期待している。

だが，これらの装置を裏づける科学は決定的に欠けている。米食品医薬品局（FDA）は 2016 年 7 月に警告を発表し，筋肉痛や不眠症，不安症の緩和その他に役立つ証拠はないと述べた。それどころか，凍傷や焼けるような痛み，目の傷害，果ては窒息を引き起こす恐れがあるという。SCIENTIFIC AMERICAN の問い合わせに答えて同局は，「FDA はいかなる全身冷却療法装置も認可しておらず，これらの装置に関する医学的主張を実証する証拠も持ち合わせていない」と付け加えた。FDA の警告は，出版ずみ文献に関する独自の非公式評価と，装置に低温条件を作り出しているガスに曝露することに伴って一般に認められる危険に基づいている。かてて加えて，全身冷却療法は費用が高い。2 分間の処置 5 回のセットで数百ドルかかる。

興ざめの肌寒さ

治療目的で全身を過冷するという考えは 1970 年代末に日本で生まれ，多発性硬化症や関節リウマチの関節痛を緩和する可能性があるとされた。その後，1990 年代に欧州で関心を集めたが，米国とオーストラリアで注目されるようになったのは 10 年ほど前からだ。実施例が広がるにつれ，対処可能とされる症状は爆発的に増えた。最近の販促資料によると，疼痛だけでなく喘息からアルツハイマー病まで様々な病気を治療できるという。

全身冷却療法の論法は，標準的な寒冷療法の根拠となっている広く認められた科学から来ている。通常の寒冷療法は急性の軟部組織損傷を氷嚢や冷水浴によって治療する。例えば足首をくじいた場合，治療の一環として氷で冷やすことが推奨されている。患部に氷を当てて 5 〜 15 分ほど冷やすと皮膚の温度が 13℃以下に下がり，傷ついた神経からの信号伝達が遅くなって痛みが和らぐことが臨床研究によって明らかになった。氷は別の作用を通じても効果をもたらすようだ。損傷部位に集まる白血球を減らすなどして炎症を抑えることが動物実験で示されていると，北アイルランドにあるアルスター大学のスポーツ医学研究者ブ

3 潜むリスク

リークリー（Chris Bleakley）はいう（炎症が長期化すると痛みが広がり，当該部位の可動範囲が狭まり，損傷周辺への血流が損なわれる）。

だが，全身冷却療法で同じ利点が実際に得られるかどうかは，よく言っても不明確だ。通常の寒冷療法とは違って，液体窒素が気化したガスを使って，患者の入ったチャンバー内の空気の温度を−130℃に下げる。この温度は氷よりもはるかに低いが，氷を身体に直接当てたほうが皮膚や脂肪を通して目的の軟部組織をうまく冷やせる。冷気の場合，皮膚の近くを渦巻いているものの押しつけられてはいないので，身体の深部まで冷やすのは難しい。

実際，冷水療法や全身冷却療法に関する既存の研究をブリークリーらが調べた2014年の解析で，皮膚表面温度と筋内温度を最も大きく下げられるのは氷嚢であることがわかった。例えば氷嚢を10分間当てると，皮膚温度は18〜26℃下がる。これに対し全身冷却療法では，3分間（使用者の安全確保のためメーカーが推奨している平均使用時間）で下げられる温度は小幅で，3〜20℃にとどまった。

筋内温度をたいして下げられないのだから，全身冷却療法は氷嚢に比べて痛みシグナルを遅らせる効果や軟部組織を冷やして炎症を抑える効果は低いだろうとブリークリーはいう。

別の研究がこの疑いを強めている。治療法の効力を評価する王道はいわゆるランダム化比較試験で，被験者を無作為に2群に分け，片方に問題の治療法を施し，他方には別の治療法を施すか何も処置せずに，結果を比較する。全身冷却療法に関してはこうしたランダム化比較試験がこれまでに4件行われている。英ポーツマス大学の運動生理学者コステロ（Joe Costello）はブリークリーらとともに，それらの研究を精査し，この治療法に有意な効果はないことを突き止めた。「筋肉痛の緩和や練習後の回復を助けることを示す証拠は不十分だ。何もしないのと変わらない」とコステロは述べる。

これら4件の試験も，コステロによるその評価も，最終決定ではない。規模が非常に小さく，被験者は合計で64人にすぎない。また被験者は4人を除

いて 20 代前半の男性なので，女性や年かさの人々に対する全身冷却療法の影響については何も言えない。

答えられていない疑問

こ れらの研究の不備は全身冷却療法に関する科学のお寒い状況を象徴している。大半の研究は被験者が「極めて少数」なうえ，対照群を設けていないなど「方法論的な欠陥」があるとブリークリーはいう。「スポーツ科学者はこの分野を習得して他の医学研究と並ぶ質に高める必要がある」。

運 動による障害のほかに全身冷却療法が対処できるとされる症状に対する効果に関しては，科学的裏づけは実質的に皆無だ。厳密なランダム化試験の例はなく，ガス化した液体窒素にさらされることが心拍数や血圧，代謝に好ましい影響を与えるかどうかについて確かなことは何もわかっていない。そうした影響が生じていれば，不安の軽減や片頭痛の治療，減量促進などに役立つ可能性も考えられるのだが。

ク ライオ USA の業務執行役マードック（Mark Murdock）は全身冷却療法に言われている多くの効用が科学的裏づけを欠いていることに反論しない。同社は疼痛や炎症の緩和と活力アップをうたって装置を販売しているが，マードックによると，装置は医学的な助力ではなく「安楽」を提供するものだ。さらに，この装置が減量を促進しうるといった主張は「ばかげている」と付け加える。また，FDA が警告を発したことを支持し，最終的には FDA が規制に乗り出して根拠のない効用の主張を抑制すべきだと考えている。

効 用に裏づけがないだけでなく，この療法に伴うリスクも明確には理解されていない。副作用に的を絞った研究がなされていないのだ。また，全身冷却療法の実施例は均一ではなく，処置時間や温度，冷気に触れないよう保護される部位などはまちまちだ。どんな条件で何度の冷気にどれだけの時間曝露されるかが安全を左右すると，米国の水球オリンピックチーム付き医師であるラオ（Naresh Rao）はいう。

それでも，見事なフリーザーの内部で一定のお勤めをして病気を治すという考え方には強い魅力がある。体験者はよい効果があったと語るが，それを裏づける科学的証拠がないということは，治療法を信じる本人の思い込みの産物にすぎない可能性をうかがわせる。プラセボ効果だ。整骨療法（全身的な治療によって伝統的な治療法を補足する方法）も手がけているラオは，身体を痛めた運動選手を治療する第一選択として全身冷却療法を選ぶことはしないものの，患者がそれを望むのなら，効果がせいぜい主観的なものにすぎなくても，試すことを支持するという。ただ，こう指摘する。「医学的に規制する必要があると思う。誰にでも実施してよいとはいえない」。例えば心臓に問題のある人や高血圧を治療せずに放っている人は避けるべきだと注意する。低温に急にさらされると，心臓発作など重大な結果を招く恐れがあるからだ。

　全身冷却療法の有効性に関して前向きなニュースを期待している研究者もいる。ラグビーチームのサンディエゴ・ブレイカーズ付きの整骨医・内科医で，大統領府のフィットネス・スポーツ・栄養審議会で科学委員を務めているロドリゲス（Rebeccah Rodriguez）はその1人だ。ロドリゲスは脳震盪（しんとう）からの回復に冷却療法装置が寄与するかどうかを評価する研究を計画している。このほか仏マルセイユの研究チームは，全身冷却療法の抗炎症効果を調べ，伝統的な非ステロイド性抗炎症薬（NSAID）を代替できるかどうかを評価する予備的な研究を進めている。

　「研究すべきことがまだ多く残っている」とブリークリーはいう。全身冷却療法の効力を測るには大規模なランダム化比較試験を行うしかない。冷徹な事実を提供して消費者を守るためにも，それが必要だ。

The Acupuncture Myth
鍼治療の神話

興味深い伝統医療ではあるが，多くの落とし穴が存在する

J. インターランディ（サイエンスライター）

掲載：SCIENTIFIC AMERICAN August 2016, 日経サイエンス 2017 年 5 月号

In 1971 then *New York Times* columnist James Reston had his appendix removed at a hospital in China. The article he wrote about his experience still reverberates today. His doctors used a standard set of injectable drugs—lidocaine and benzocaine—to anesthetize him before surgery, he explained. But they controlled his postoperative pain with something quite different: a Chinese medical practice known as acupuncture, which involved sticking tiny needles into his skin at very specific locations and gently twisting them. According to Reston, it worked.

Readers back home were fascinated. In a rush of excitement over this new, exotic knowledge, the original story was quickly jumbled. Before long, it was commonly believed that the Chinese doctors had used acupuncture not just after Reston's appendectomy but as anesthesia for the surgery itself. Interest in acupuncture soared in the U.S. and has remained high ever since.

But it turned out that acupuncture as Reston described it was not the enduring bit of ancient Chinese wisdom enthusiasts supposed. In fact, the procedure had been written off as superstition back in the 1600s and abandoned altogether in favor of a more science-based approach to healing by the 1800s. Chinese Communist

Vocabulary

appendix 虫垂
reverberate 反響する, 波紋を広げる
lidocaine リドカイン
benzocaine ベンゾカイン
anesthetize 麻酔する

acupuncture 鍼（はり）

jumble 混乱させる

appendectomy 虫垂切除術
anesthesia 麻酔
soar 高まる

enduring 不朽の
enthusiast 熱狂者
write off 〜とみなす
superstition 迷信

119

Party leader Mao Zedong had only revived acupuncture in the 1950s as part of his initiative to convince the Chinese people that their government had a plan for keeping them healthy despite a woeful dearth of financial and medical resources.

Vocabulary

woeful ひどい, 悲惨な
dearth 欠乏

Even more impressive than how well Mao's campaign worked in China at the time is how well it is working in the U.S. today. Every year hundreds of thousands of Americans undergo acupuncture for conditions ranging from pain to post-traumatic stress disorder, and the federal government spends tens of millions of dollars to study the protocol.

post-traumatic stress disorder 心的外傷後ストレス障害 (PTSD)
protocol 慣例, 手順

So far that research has been disappointing. Studies have found no meaningful difference between acupuncture and a wide range of sham treatments. Whether investigators penetrate the skin or not, use needles or toothpicks, target the particular locations on the body cited by acupuncturists or random ones, the same proportion of patients experience more or less the same degree of pain relief (the most common condition for which acupuncture is administered and the most well researched). "We have no evidence that [acupuncture] is anything more than theatrical placebo," says Harriet Hall, a retired family physician and U.S. Air Force flight surgeon who has studied, and long been a critic of, alternative medicine.

sham にせの, ごまかしの

cite 例として挙げる

relief 軽減
administer 適用する, 施す

placebo プラセボ, 偽薬
flight surgeon 航空軍医
alternative medicine 代替医療

But the news is not all bad. In the process of putting acupuncture to the test, scientists have gained insights that could lead to the development of new and urgently needed methods for treating pain.

SMALL EFFECTS

Acupuncture is based on the concept of qi (pronounced "chi"), a life force or energy that practitioners say flows through the body along 20 distinct routes

qi 気

called meridians. Blocked meridians are believed to cause illness by disrupting the flow of qi. Inserting acupuncture needles at specific points along specific meridians is thought to clear those blockages and restore qi's natural flow, which in turn restores patients to health. Scientists have long understood that qi is not a legitimate biological entity; many studies have shown that the effects of acupuncture are the same whether needles are placed along the meridians or at random locations around the body. But the acupuncture proponents among them have argued that acupuncture itself might still work, albeit by an as yet unknown mechanism.

Some of the best support for this contention came in 2012, when researchers at Memorial Sloan Kettering Cancer Center and their colleagues published a meta-analysis of 29 studies involving nearly 18,000 patients, which found that traditional acupuncture produced a somewhat greater reduction in pain than placebo or sham acupuncture. The finding was widely touted as the first clear proof that acupuncture actually works. But critics have dismantled that interpretation. For one thing, they point out, acupuncture studies are extremely difficult to double-blind—a methodological approach in which neither the researchers nor patients know who is receiving the treatment under investigation and who is receiving the placebo or sham. The researchers knew which patients were and were not getting real acupuncture, and that awareness almost certainly biased their results. In addition, although statisticians detected a difference in pain relief between treatment and placebo, the effect may have been lost on patients. "What [the study authors] are arguing is that a change of 5 on a 0–100 pain scale … is noticeable by patients," David Gorski, a surgical oncologist at the Wayne State University School of Medicine, observed in a blog post. "It's probably not."

Vocabulary

meridian 経絡

legitimate 正当な, 合理的な
entity 実体

albeit にもかかわらず, ではあるが

contention 主張

meta-analysis メタ解析

tout ほめそやす

dismantle 飾りつけをはずす, 実際を暴く
double-blind 二重盲検法で調べる
▶ 111ページ
Technical Terms

statistician 統計学者, 統計の専門家

surgical oncologist 腫瘍外科医

The lack of scientific support for acupuncture has not curbed enthusiasm for the practice. Blue-chip medical centers such as the Mayo Clinic and Massachusetts General Hospital now have dedicated acupuncturists on staff. Health insurance programs are starting to cover acupuncture to a limited extent, and individual consumers who cannot get insurance to foot the bill are collectively shelling out millions from their own pockets. Nor have the findings stopped the flow of government money into acupuncture programs, which has totaled more than $73 million since 2008. In that time, Mass General has received $26 million in such funding from the Department of Health and Human Services, largely for studies that scan the brains of people being treated with acupuncture or thinking about being treated with acupuncture. And the Department of Defense has awarded more than $12 million in acupuncture contracts and grants.

Part of that continued investment could have to do with patient demand. But there are other justifications. Josephine Briggs, director of the National Center for Complementary and Integrative Health (the NIH division responsible for all alternative medicine research), acknowledges that the balance of evidence points to a placebo effect for acupuncture. Yet in her view, there is still good reason to study the procedure. "It isn't implausible that the effect of a lot of needles may change central pain processing in some concrete way," she says. Just as the finding that tea made from willow bark could alleviate headaches led scientists to the discovery of salicylic acid—which in turn led to the invention of aspirin—many acupuncture researchers think that their work might lead to a treatment for pain that is more effective than acupuncture. Their goal, in other words, is not to justify acupuncture per se but to find out if a mechanism of some kind can explain those very small effects and, if so, whether that mechanism can be exploited to produce a viable treatment for pain.

A POSSIBLE MECHANISM

With this goal in mind, scientists have been studying a roster of potential biological pathways by which needling might relieve pain. The most successful of these efforts has centered on adenosine, a chemical believed to ease pain by reducing inflammation. A 2010 mouse study found that acupuncture needles triggered a release of adenosine from the surrounding cells into the extracellular fluid that diminished the amount of pain the rodents experienced. The mice had been injected with a chemical that made them especially sensitive to heat and touch. The researchers reported a 24-fold increase in adenosine concentration in the blood of the animals after acupuncture, which corresponded to a two-thirds reduction in discomfort, as revealed by how quickly they recoiled from heat and touch. Injecting the mice with compounds similar to adenosine had the same effect as acupuncture needling. And injecting compounds that slowed the removal of adenosine from the body boosted the effects of acupuncture by making more adenosine available to the surrounding tissue for longer periods. Two years later a different group of researchers went on to show that an injection of PAP, an enzyme that breaks other compounds in the body down into adenosine, could relieve pain for an extended chunk of time by increasing the amount of adenosine in the surrounding tissue. They dubbed that experimental procedure "PAPupuncture."

Both sets of findings have excited researchers—and for good reason. The current options for treating pain are limited and rely mostly on manipulating the body's natural pain-management system, known as the opioid system. Opioid-based painkillers are problematic for several reasons. Not only does their efficacy tend to wane over time, but they have been linked to an epidemic of addiction and overdose deaths across the U.S.—so much so that the Centers for Disease Control and Prevention has recently advised doctors to seriously restrict their use. The

Vocabulary

roster 一覧表

adenosine アデノシン
inflammation 炎症

extracellular fluid 細胞外液
rodent 齧歯（げっし）類

discomfort 苦痛, 不快
recoil 後ずさりする

removal 除去

enzyme 酵素

opioid system オピオイド系
wane 薄れる, 弱まる
epidemic 流行, 蔓延
addiction 依存
overdose 過剰投与
Centers for Disease Control and Prevention 米疾病対策センター

available nonopioid pain treatments are few; many of them require multiple injections or catheterization to work; and they often come with side effects, such as impaired movement. Adenosine offers an entirely new mechanism to exploit for potential treatments—one that may come with fewer side effects and less potential for addiction. What is more, adenosine can be made to circulate in the body for prolonged stretches. Pharmaceutical companies are actively investigating adenosine-related compounds as potential drugs.

But however promising adenosine may be as a treatment, the findings from this research do not prove that acupuncture itself "works." For one thing, the researchers did not show that the release of adenosine was specific to acupuncture. Acupuncture needles might cause adenosine to flood the surrounding tissue, but so might a hard pinch, or applied pressure, or any number of other physical insults. In fact, both of the studies found that when adenosine was turned on in mouse tissue by other mechanisms, the pain response was equal to or better than the response generated by acupuncture. For another thing, the study results offered no support for the use of acupuncture to treat any of the other conditions for which the procedure is often advertised. A localized adenosine response may mitigate localized pain. That does not mean it can also cure insomnia or infertility.

It may well be that the reams of research scientists have done on acupuncture have lit the path toward improved understanding of—and eventually better treatments for—intractable pain. But it may also be time to take whatever bread crumbs have been laid out by that work and move on.

Vocabulary

catheterization カテーテル法による投与

pinch つねること
insult 傷害

insomnia 不眠症
infertility 不妊

reams of たくさんの

intractable 手に負えない, 治りにくい
bread crumbs パンくず

1971年，*New York Times*紙のコラムニストだったレストン（James Reston）は中国の病院で虫垂切除手術を受けた。この経験について述べた彼の記事は現在もなお波紋を広げている。彼の説明では，主治医は手術前の麻酔に標準的な注射剤（リドカインとベンゾカイン）を用いたが，術後の痛みを抑えるためにまったく異なる方法を採用した。鍼（はり）として知られる中国の伝統医療で，特定のツボに小さな針を刺して静かに動かす。レストンによると，これが効いた。

米国の読者は夢中になった。この目新しく風変わりな治療法に興奮したあまり話に尾ひれがつき，中国の医師たちは単にレストンの術後に鍼を用いただけでなく虫垂切除術そのものの麻酔に使ったのだと信じられるようになった。鍼に関する米国内の関心は燃え上がり，以来ずっと高いままだ。

だが熱狂者の想像に反して，レストンが述べた鍼治療が中国古来の不朽の知恵というわけではないことがわかった。それどころか，17世紀には迷信と見なされ，19世紀になると科学に基づく他の方法が好まれるなか完全に廃れていた。ようやく復活したのは1950年代，中国共産党の指導者だった毛沢東による。資金も医療資源も乏しい窮状にあって，国民の健康を維持する手段として鍼が有効だとして推進した。

毛沢東のキャンペーンは当時の中国で効果を上げたが，より印象深いのは現在の米国で同様のキャンペーンが成功していることだ。毎年何十万人もの人々が疼痛から心的外傷後ストレス障害（PTSD）まで様々な症状について鍼治療を受け，政府は何千万ドルもの資金を鍼治療の研究に費やしている。

研究成果はこれまでのところ期待はずれだ。鍼治療と様々なインチキ療法の間に有意な差は見つかっていない。皮膚に穴を開けても開けなくても，針を使っても爪楊枝を使っても，鍼療法師がいう特定のツボを狙ってもランダムな場所を選んでも，患者のうち同じ割合の人が多かれ少なかれ痛みが和らぐのを経験する（鍼治療が施される最も一般的な目的は痛みの緩和で，研究も最も進んでいる）。「鍼治療が芝居じみたプラセボ（偽薬）以上のものである証拠はない」と，元空軍の航空軍医ホール（Harriet Hall）はいう。彼女は長年にわたり代替医療を調べ，批判してきた。

3　潜むリスク

だが，悪い話ばかりではない。科学者は鍼治療を検証するなかで，新たな疼痛治療法の開発につながりそうな知見も手に入れてきた。

小さな効果

鍼治療は漢方でいう身体のエネルギー「気」の概念に基づいている。気は身体のなかを 20 の「経絡」に沿って流れているとされ，経絡がブロックされると気の流れが乱れて病気を起こすのだと信じられている。経絡に沿った特定のツボに針を刺すと経絡の閉塞が解け，気の自然な流れが復活して，健康を取り戻せるという。科学者たちはずっと以前から，気が生物学的に合理的な実体ではないことを理解している。針を経絡沿いに刺してもランダムな場所に刺しても鍼治療の効果が同じであることを多くの研究が示している。だが鍼治療の推進者は，メカニズムは不明ながら，鍼治療自体は効果があるのだと反論してきた。

この主張を支えるこれまでで最善の根拠は，スローン・ケタリング記念がんセンターの研究者らが 2012 年に発表したメタ解析の結果だ。合計で 1 万 8000 人近い患者を対象にした 29 件の研究結果を解析したもので，伝統的な鍼治療がプラセボや偽の鍼治療と比べ痛みの低減にやや大きな効果を上げたという。この発見は鍼治療が実際に効くことを示す初の明確な証拠としてもてはやされたが，批判派が問題点を暴いた。まず，鍼治療の調査研究は二重盲検法が極めて難しい。誰が調査対象の治療法を施され，誰が対照のプラセボや偽治療を施されているのかを，研究者も患者も知らない状態で行うのが二重盲検法だ。誰が鍼治療を受けているかを研究者が知っていると，それによって結果にほぼ確実にバイアスが生じる。さらに，鍼治療とプラセボで痛みの緩和に統計的な違いが認められたものの，患者によってはこの効果はなかった可能性がある。「患者は痛みが 100 段階スケールにして 5 段階相当だけ変化したのを認めたと研究論文の著者は主張しているが，おそらくそうではない」とウェイン州立大学（ミシガン州）の腫瘍外科医ゴースキー（David Gorski）はあるブログ投稿記事で述べている。

科学的根拠が欠けていても，鍼治療に対する熱狂は冷めていない。メイヨークリニックやマサチューセッツ総合病院といった一流の病院が鍼治療専門のスタッフを置いている。健康保険は限定的ではあるが鍼治療をカバーし始め，保険に入っていない人も自費で総額数百万ドルを鍼治療に支払っている。政府が

鍼治療研究にかける費用も衰えず，2008年以来の総額で7300万ドルを超えた。マサチューセッツ総合病院は米保健福祉省（HHS）から2600万ドルの研究費を得て，鍼治療を受けている人や鍼治療を受けていると想像している人の脳画像を撮影して調べている。また米国防総省は鍼治療の実施費用や研究助成金として，これまでに1200万ドル以上を支出した。

　こうした投資が続いているのは一部には患者の求めによるのだろうが，別の正当な理由もある。米国立補完統合衛生センター〔米国立衛生研究所（NIH）の一部門で代替医療の研究を担当〕の所長ブリッグス（Josephine Briggs）は，鍼治療がプラセボ効果であることを示す科学的証拠が優勢であると認めつつ，この処置を研究する十分な理由があると考えている。「多数の針の効果が中枢神経系の痛み信号処理を具体的に変える可能性は，考えられないことではない」という。柳の樹皮から作ったお茶で頭痛が和らぐという発見がサリチル酸の発見とアスピリンの発明につながったように，多くの鍼治療研究者は自分たちの研究が鍼治療よりも効果的な疼痛治療法につながるかもしれないと考えている。つまり研究の目的は鍼治療そのものを立証することではなく，鍼治療の非常に小さな効果を説明できるメカニズムの有無を調べ，そうしたメカニズムが存在するならそれを利用して有効な疼痛治療ができるかどうかを見極めることだ。

考えうるメカニズム

　この目的を念頭に，様々な生物学的経路が研究されている。最も有望そうなのは，炎症を抑えることで痛みを和らげると考えられているアデノシンという物質に着目した研究だ。マウスで行われた2010年の研究は，鍼治療がきっかけとなって周辺の細胞から細胞外液にアデノシンが放出され，痛みが軽減したことを示した。実験ではあらかじめ，マウスにある化学物質を注射し，熱と接触刺激に特に敏感にしておいた。鍼を打った後は血液中のアデノシン濃度が24倍に増え，痛みの辛さが2/3に軽減した（熱や接触からどれだけ素早く身を引くかで評価）。また，アデノシンに似た物質をマウスに注射すると，鍼を打った場合と同じ効果が生じた。さらに，アデノシンが体内から排出されるのを遅らせる物質を注射すると，周辺組織にアデノシンが長く残り，鍼治療の効果が高まった。2年後には別の研究グループが，体内の別の物質を分解してアデノシンに変えるPAPという酵素を注射すると，周辺組織でアデノシンの量が増え長期にわたって

痛みが和らぐことを示した。研究チームはこの実験的処置をacupuncture（鍼治療）ならぬ「PAPupuncture」と名づけた。

　これらの発見に研究者たちは当然ながら興奮した。現在の疼痛治療法は限られており，多くは人体の痛み管理システムである「オピオイド系」を操作することによっている。オピオイドに基づく鎮痛剤はいくつかの理由で問題含みだ。効果が時間とともに薄れるだけでなく，米国では依存者の増加や過剰投与による死亡が問題になっている。米疾病対策センター（CDC）は最近，オピオイド系鎮痛剤の利用を厳しく制限するよう医師に勧告した。非オピオイド系の鎮痛剤は数少ないうえ，複数回の注射かカテーテル法による投与が必要なものが多い。また運動障害など様々な副作用を伴う。これに対しアデノシンは疼痛治療につながりうるまったく新しいメカニズムを提供しており，副作用が少なく依存の心配も小さい可能性がある。さらに，より長時間にわたって体内を循環するよう調整できる見込みがある。医薬品各社はアデノシン関連物質の可能性を積極的に研究している。

　しかし，アデノシンがいかに有望とはいえ，この研究は鍼治療そのものが"効く"という証明ではない。まず，アデノシンの放出が鍼治療に特有であると示してはいない。針が周辺組織にアデノシンの増加を引き起こしているのかもしれないが，強くつねったり圧力を加えたり，その他の物理的な傷害でも同じことが起こるかもしれない。実際どちらの研究も，鍼治療以外の理由でアデノシンがマウスの組織で増えた場合に，鍼治療によるのと同等以上の痛み低減効果が生じることを見いだした。さらに，この研究結果は鍼治療が効くとされる痛み以外の症状に対する有効性について何も示していない。アデノシンの局所的な反応は局所的な痛みを緩和するのかもしれないが，それは不眠症や不妊を治療できることを意味しない。

　多くの研究者が鍼治療を研究して難治性疼痛に関する理解（と最終的にはよりよい治療法）につながる道筋を明らかにしたのは結構なことだろう。だが，その仕事によってどんな道が開けたのかを一度立ち止まって考えてから前に進む段階を迎えたともいえるだろう。

Overreaction
過剰反応のアレルギー検査

不正確な検査のせいで，多くの子供が食物アレルギーと誤診されている

E. R. シェル（ジャーナリスト）

掲載：SCIENTIFIC AMERICAN November 2015, 日経サイエンス 2016 年 5 月号

Just a few years ago a 15-month-old girl—her stomach, arms and legs swollen and her hands and feet crusted in weeping, yellow scales—was rushed to the emergency room at the University of Texas Southwestern Medical Center in Dallas. Laboratory tests indicated a host of nutrition problems.

The child's mother, during the previous year, had told doctors that standard infant formula seemed to provoke vomiting and a rash. The mother and her pediatrician assumed the girl was allergic to the formula and switched her to goat's milk. Symptoms persisted, though, and the baby was switched again, to coconut milk and rice syrup. At 13 months, the pediatrician noted yet another red, swollen rash and ordered an allergy test, the child's first. The test identified coconut as a so-called high-reaction class, and coconut milk was removed from her diet. Reduced to a diet of rice milk, the child's symptoms worsened.

In the ER, doctors determined the girl suffered from kwashiorkor, a nutritional disorder rarely seen in the developed world. She was fed intravenously and evaluated by a team that included pediatric allergist J. Andrew Bird, who used more sophisticated methods to test her response to coconut and cow's milk, wheat, soy, egg white, fish,

Vocabulary
stomach 腹部
weeping 滲出性の, じくじくする

formula 調合乳
vomiting 嘔吐
rash 発疹
pediatrician 小児科医

kwashiorkor クワシオルコル（栄養失調の一種）
intravenously 静脈を通じて

shrimp, green beans and potatoes. To her mother's astonishment, the toddler showed no adverse reaction to any of them. After a few days of steady nourishment and a course of antibiotics to clear her skin of various infections, she was released from the hospital into a life free of food restrictions. (Her digestive upsets appeared to be caused by a variety of common ailments that would have almost certainly cleared on their own.)

The problem was not in the baby but in the tests. Common skin-prick tests, in which a person is scratched by a needle coated with proteins from a suspect food, produce signs of irritation 50 to 60 percent of the time even when the person is not actually allergic. "When you apply the wrong test, as was the case here, you end up with false positives," says Bird, who co-authored a paper describing the Dallas case in 2013 in the journal *Pediatrics*. And you end up with a lot of people scared to eat foods that would do them no harm. Bird has said that he and a team of researchers found that 112 of 126 children who were diagnosed with multiple food allergies tolerated at least one of the foods they were cautioned might kill them.

Kari Nadeau, director of the Sean N. Parker Center for Allergy Research at Stanford University, says that many pediatricians and family physicians are not aware of these testing flaws. "When it comes to diagnosis, we've been in the same place for about 20 years," she observes. To move forward, Nadeau and other researchers are developing more advanced and easily used methods.

Food allergies are real and can be deadly, but mistakenly slapping an allergy label on a patient can be a big problem as well. First, it does not solve the person's troubles. Second, a diagnosis of allergies comes with a high price: a few years ago Ruchi S. Gupta, a pediatric allergist affiliated with the Northwestern University Feinberg

School of Medicine, estimated the annual cost of food allergy at nearly $25 billion, or roughly $4,184 per child, with some of that attributed to medical costs but even more to a decline in parents' work productivity.

There is a mental health price as well: children who believe they have a food allergy tend to report higher levels of stress and anxiety, as do their parents. Every sleepover, picnic and airplane ride comes fraught with worry that one's child is just a peanut away from an emergency room visit or worse. Parents and children must be ever armed with an injectable medicine that can stave off a severe allergic reaction. The prospect of a lifetime of this vigilance can weigh heavily on parents, some of whom go so far as to buy peanut-sniffing dogs or to homeschool their children to protect them both from exposure to the offending food and from the stigmatization of the allergy itself.

Pediatric allergist John Lee, director of the Food Allergy Program at Boston Children's Hospital, has heard more than his share of horror stories. "Food allergies can be terribly isolating for a kid," he says. "One parent told me his child was forced to sit all alone on a stage during lunch period. And siblings can feel resentful because in many cases parents don't feel they can take family vacations or even eat dinner in a restaurant."

Diagnosing a food allergy usually begins with a patient history and the skin-prick test. If the scratch does not provoke a raised bump surrounded by a circle of red itchiness, the patient almost certainly is not allergic to the material. But positive tests can be harder to interpret because skin irritation does not necessarily reflect a true allergy, which is a hypersensitivity of the immune system that extends through the body. In a real allergy, immune components such as IgE antibodies in the blood are stimulated by an allergen. The antibody binds to immune cells called mast cells, which then triggers release of a cascade of chemicals that produce all kinds of inflammation and irritation. But levels of allergen-specific antibodies in the blood are quite low even in allergic people, so running a simple blood test is not an answer, either.

The diagnostic "gold standard" for food allergy is a placebo-controlled test. A potential irritant is eaten, and the body's response (a rash, say, or swelling) is compared with what happens after eating something that looks like the irritant but is benign. For example, a patient who might be allergic to eggs is given a tiny amount of egg baked into a cake, along with a taste of egg-free cake. Ideally, the test is double-blind, meaning that neither the patient nor the allergist knows which cake contains egg. The accuracy rate of these tests, for both positive and negative results, is about 95 percent, according to Lee.

Unfortunately, this procedure is tricky, time-consuming, expensive and relatively uncommon; experts agree that few allergy sufferers have access to it.

James Baker, who is a physician and immunologist and CEO of the nonprofit Food Allergy Research & Education (FARE), says his organization is tackling this problem by setting up 40 centers around the country to administer food challenges with all the necessary precautions. "You have to be prepared to treat or transport people to the emergency room if they react," he asserts.

Scientists are also looking for something easier to use. One promising newcomer to the diagnostic arsenal is the basophil-activation test (BAT). Basophils, a type of white blood cell, excrete histamines and other inflammatory chemicals in reaction to a perceived threat—such as an allergen. Nadeau and her colleagues have designed and patented a test that involves mixing just one drop of blood with the potential allergen and measuring the reaction in basophils. In pilot studies, the procedure diagnosed allergies with 95 percent accuracy in both children and adults, a rate similar to that of food-challenge tests.

BAT is still in the research phase and requires more studies with a larger, more varied population, but another approach—allergen-component testing—has already been approved by the U.S. Food and Drug Administration for peanut allergies. Lynda Schneider, a pediatric allergist and director of the Allergy Program at Boston Children's Hospital, says that some children have a mild sensitivity—but not a full-blown allergy—to one protein in peanuts. Rather than testing them with crude mixtures of lots of proteins found in nuts, Schneider's component tests isolate specific proteins and then challenge the patient with those. By sorting out which protein is prompting the negative reaction, physicians can determine with a high degree of accuracy whether the patient is truly allergic to peanuts.

Schneider wants to get beyond diagnosis and into treatment. Omalizumab is a monoclonal antibody that binds to IgE antibodies and prevents them from glomming on to mast cells, which triggers the allergic cascade. In a recent study, Schneider and her colleagues administered this so-called anti-IgE drug over the course of 20 weeks to 13 children who were known to have peanut allergies while giving them a gradually larger dose of peanuts. During the anti-IgE phase, none of the children developed an allergic reaction to peanuts, although two did have a recurrence once the anti-IgE regime ended. "The anti-IgE allowed their system to go through a desensitization process," Schneider says.

Kids who are allergic to milk and eggs can be gradually desensitized by heating these foods for 30 minutes or so, Bird has found. The heat changes the shape of these proteins, which vastly reduces their tendency to provoke

過剰反応のアレルギー検査

Vocabulary

allergen-component testing アレルゲンコンポーネント検査
U.S. Food and Drug Administration 米食品医薬品局

sort out 選り分ける, 見つけ出す

omalizumab オマリズマブ
monoclonal antibody モノクローナル抗体
▶ 72 ページ Technical Terms
glom on to くっつく

dose 投与量

recurrence 再発, ぶり返し

desensitization 減感作

133

allergies. This is not a home remedy, and it is done under medical supervision, but studies of kids who are fed small amounts of heated egg or milk show the children are far more likely to acquire a tolerance to these foods over time—that is, more likely to outgrow the allergy. A study called Learning Early About Peanut Allergy (LEAP) showed that exposing children to tiny amounts of peanut products early in their life dramatically reduced the incidence of allergy.

Scott H. Sicherer, a professor of pediatrics, allergy and immunology at the Icahn School of Medicine at Mount Sinai, takes the early desensitization idea a step further. He suggests children can best avoid food allergies if they eat a wide variety of foods at an early age, run in the open air and "play in the dirt." A little less protection from the world, he says, may be the best protection from allergies.

Vocabulary

home remedy 家庭での療法

tolerance 耐性

outgrow 脱する

つい2〜3年前，生後15カ月の女児がダラスにあるテキサス大学南西メディカルセンターの救急救命室に担ぎ込まれた。腹部と四肢が腫れ，手足に滲出性の黄色いかさぶたができている。検査の結果，様々な栄養素の欠乏が示された。

その前年，母親はこの子を医師に診せ，通常の調合乳を与えると吐き戻してしまい，発疹が出るようだと訴えた。小児科医も調合乳に対するアレルギーを疑い，山羊乳に切り替えた。だが症状は続き，今度はココナツミルクとライスシロップにした。この子が13カ月齢になったとき，小児科医は別の赤い発疹に気づいてアレルギー検査を初めて行った。この結果，ココナツがいわゆる高反応クラスと判定され，食事をライスミルクだけにしたが，症状はさらに悪化した。

救急救命室の医師たちは，この女児が先進国ではまれな「クワシオルコル」という栄養失調であると結論づけた。静脈栄養補給に切り替えられ，小児アレルギー専門医バード（J. Andrew Bird）らがさらに精巧な方法を用いてココナツミルクと牛乳，小麦，大豆，卵白，魚，エビ，サヤインゲン，ジャガイモに対する反応を調べた。母親が驚いたことに，この子はこれらのいずれにも有害な反応を示さなかった。2〜3日の栄養補給と，抗生物質による皮膚感染菌の一掃処置の後，この子は退院し，食事制限なしに暮らしている（この子の不調は一般的な軽い病気が様々に重なって生じたものらしく，自然治癒したとみられる）。

問題は女児ではなく検査のほうにあった。よく行われる「皮膚プリックテスト」は，原因とおぼしき食物が含んでいるタンパク質をつけた針で皮膚を引っ掻いて反応を見るもので，アレルギーでない場合でも検査の50〜60%で炎症のサインが生じる。「今回のように検査が不適切だと，誤ってアレルギーと判定される」とバードはいう（彼はダラスの例を共著論文にまとめ，*Pediatrics*誌に報告した）。そして，多くの人が無害な食物を避ける結果になってしまう。複数の食物アレルギーと診断された126人の子供のうち112人までが，食べると命にかかわると警告された食物の少なくとも1つに実は耐性があったことが判明したという。

スタンフォード大学アレルギー研究センター所長のナドー（Kari Nadeau）は，小児科医やかかりつけ医の多くがこうした検査の欠陥を知らないという。「ことアレルギーの診断に関しては20年前からまるで進歩していない」とみて，

ナドーらはより進んだ使いやすい検査法の開発を進めている。

食物アレルギーは確かに存在するし致死的な場合もあるが，誤ってアレルギーのレッテルを貼るのも同様に大きな問題となりうる。まず，それでは患者のトラブルは解決しない。第2に，アレルギーの診断は高くつく。ノースウェスタン大学医学部の小児アレルギー専門医グプタ（Ruchi S. Gupta）は数年前，食物アレルギーに伴う米国のコストは年間250億ドル近く，小児患者1人あたり年間約4184ドルになると推算した。この一部は医療費だが，子供の世話や通院などに伴う両親の労働生産性の低下による部分が大きい。

精神衛生上の損失もある。自分が食物アレルギーだと思っている子はそうでない子に比べ，ストレスや不安を感じる傾向が強い。外泊やピクニック，飛行機旅行はみな，何かを食べて病院に担ぎ込まれるかもしれないという恐怖と隣り合わせだ。親も子も，激烈なアレルギー反応を抑えるための注射剤を常に携帯しなければならない。そうした警戒を一生続けねばならないという見通しが両親に強くのしかかり，なかにはピーナツを嗅ぎ出す犬を買い求めた人や，わが子を自宅で教育している親もいる。学校で厄介な食物にさらされるのを避けるほか，アレルギーで外聞の悪い思いをするのを回避するためだ。

ボストン小児病院で食物アレルギープログラムを率いているリー（John Lee）はぞっとする話をいやというほど聞いてきた。「食物アレルギーは子供をひどく孤立させる」と彼はいう。「ある父親の話では，子供が学校で給食の間ずっと教壇に1人で座らされた。また，家族旅行はもちろんレストランで一緒に食事することも無理だと考える親が多いので，アレルギーの子の兄弟姉妹はそうした楽しみを奪われたことを恨みに思いがちだ」。

食物アレルギーの診断はふつう，問診と皮膚プリックテストから始まる。テストで赤い発疹が生じなかった場合，アレルギーでないことはほぼ確実だ。だが陽性の結果は解釈が難しい。皮膚の炎症がアレルギーの結果とは限らないからだ。アレルギーは全身の免疫系が過敏になった状態であり，血液中の抗体IgE（免疫グロブリンE）などの免疫分子がアレルゲンによって刺激されて起こる。この抗体がマスト細胞（顆粒細胞）という免疫細胞に結合し，炎症を引き起こす一連の

化学物質が放出される。だが，アレルギー患者でも問題のアレルゲンに反応する抗体の血中濃度は極めて低いので，簡単な血液検査ではわからない。

食物アレルギーを診断する最高の判定基準はプラセボ対照試験だ。刺激物と思われるものと，それとよく似ているが無害なものを食べたときとで，身体の反応（発疹や腫脹など）を比較する。例えば鶏卵アレルギーが疑われる場合なら，少量の卵を焼き込んだケーキと卵を使っていないケーキを食べて比較する。できれば，患者も医師もどちらのケーキが卵を含んでいるかを知らずに実施する二重盲検法が望ましい。リーによると，このテストの正確さは，陽性判定と陰性判定のどちらについても，約95％に達する。

残念ながら，この検査は手間と時間がかかり，費用もかさむので，あまり使われていない。アレルギー患者でこのテストを受けられる人は少ないと専門家も認めている。

そこで，内科医・免疫学者のベーカー（James Baker）が代表を務める非営利団体FARE（食物アレルギー研究・教育）は，必要な予防措置のもとで食物抗原投与ができる検査センターを全米40カ所に整備中だ。「反応が実際に起こった場合，治療するか救急救命室に搬送する必要がある」からだ。

より手軽な方法も追求されており，有望な新診断法に「好塩基球活性化試験（BAT）」がある。好塩基球は白血球の一種で，アレルゲンなど外来の脅威に反応してヒスタミンをはじめとする炎症性の物質を分泌する。ナドーらは，アレルゲンと思われる物質をたった1滴の血液に混ぜて好塩基球の反応を測る検査法を開発して特許を取得した。初期の研究では，子供と大人の両方についてアレルギーを95％の確度で診断できた。この確度は食物抗原投与テストと同じだ。

BATはまだ研究段階で，今後さらに大人数の集団で試す必要があるが，「アレルゲンコンポーネント検査」という別の方法はピーナツアレルギーの診断用にすでに米食品医薬品局（FDA）の認可を得た。ボストン小児病院でアレルギープログラムを率いるシュナイダー（Lynda Schneider）は，ピーナツが含むあるタンパク質1種に対して，本格的なアレルギーではないが，やや敏感な子供がいる

という。コンポーネント検査では，ナッツ類に見られる様々なタンパク質の混合物ではなく，各タンパク質を分離したうえで投与して反応を見る。どれが有害な反応を引き起こすかを特定することで，その患者が本当にピーナツアレルギーなのか高精度で決定できる。

シュナイダーは診断から治療に歩を進めたいと考えている。「オマリズマブ」という薬は IgE 抗体に結合するモノクローナル抗体で，IgE がマスト細胞にくっつくのを妨げてアレルギー反応が始まらないようにする。いわゆる抗 IgE 抗体医薬だ。シュナイダーらは最近の研究で，ピーナツアレルギーの子供 13 人に 20 週間にわたってオマリズマブを投与しながら，ピーナツを少しずつ量を増やして与えた。抗 IgE 抗体を投与している間は，アレルギー反応を起こした子はいなかった（ただ，投与をやめた後に再発した子が 2 人いた）。「抗 IgE 抗体を用いることで減感作療法が可能になった」とシュナイダーはいう。

牛乳や卵にアレルギーを起こす子供の場合，これらの食品を 30 分ほど加熱して与えることで過敏な反応が徐々に収まることをバードは発見した。食品に含まれるタンパク質の形が加熱によって変わり，アレルギーを引き起こす傾向が著しく弱まる。これは専門医の監督のもとに行われた実験で，家庭での民間療法ではないが，アレルギーの子供に加熱した卵や牛乳を少量与え続けるとついには耐性を獲得することが示されている。つまりアレルギーを脱する可能性が高まるのだ。LEAP（Learning Early About Peanut Allergy）という研究では，子供たちに早くから少量のピーナツを食べさせるとアレルギーの発症が劇的に減ることが示された。

マウントサイナイ医科大学で小児科学とアレルギー，免疫学の教授を務めるサイカラー（Scott H. Sicherer）は，この考え方をさらに一歩進めて，食物アレルギーにならないための最良の方法は幼いころから様々な種類の食物を食べ，屋外で走り回り，「泥んこになって遊ぶことだ」という。外界に対する防護を少し緩めることが，アレルギーを防ぐ最強の防護となるのだろう。

Deadly Drug Combinations
危うい薬の飲み合わせ

コンピューター利用と遺伝子解析で悪影響を予測できる可能性も

J. ワプナー（サイエンスライター）

掲載：SCIENTIFIC AMERICAN October 2015, 日経サイエンス 2016 年 4 月号

Tucking a spreadsheet in among the toiletries in the bathroom cabinet might seem a bit odd, but for 76-year-old Barbara Pines, it is the easiest way to keep track of all the prescription medications, over-the-counter pills and supplements that she and her husband take. The document lists 20 drugs—along with the strength, number of times taken and purpose. "I print this schedule and take it to any new doctor we go to," she says.

Pines is among the 40 percent of Americans who are 65 years of age or older and take more than five prescription drugs. Although older individuals account for the majority of prescription drug users, they are hardly alone. More than four billion prescriptions were filled at U.S. pharmacies in 2014—an average of nearly 13 per citizen at that time.

The need to take multiple drugs poses a special risk that too often goes unrecognized by doctors and patients: certain combinations of medicines (prescription or otherwise) cause side effects that do not arise when the individual substances are taken alone. Studies published over the past two decades suggest that such "drug interactions" cause more than 30 percent of side effects from medications. Unfortunately, pharmaceutical manufacturers cannot always predict when a new agent will mix badly

Vocabulary

toiletries 洗面用品
keep track of ～の動静を把握する
prescription medication 処方薬

prescription drug 処方薬

drug interactions 薬物相互作用

with other medicines—not to mention supplements or foods—and so unexpected deaths are sometimes the first sign of danger.

Not all side effects are lethal, but the widespread danger from drug interactions is prompting new efforts to prevent people from taking risky combinations. Much of this work depends on finding informative patterns in huge masses of disparate data.

PILLS AND PATHWAYS

Drug interactions typically occur when the body breaks down, or metabolizes, medicines. Common trouble spots are the intestine, where ingested drugs are released into the bloodstream, and the liver, where most drugs get degraded.

In the liver, breaking down drugs is primarily the task of a family of enzymes called cytochrome P450. In fact, just six of the approximately 50 enzymes in this family digest 90 percent of all known medications. Problems can arise when two drugs require processing by the same cytochrome. If one of the drugs blocks this enzyme's activity, then too little of the second drug will be degraded and too much will remain in the bloodstream. If, on the other hand, the cytochrome gets a boost from the first drug, then the second drug will have a diminished effect because the enzyme will remove it from the body too quickly. Drugs can also bind to one another in the intestinal tract before ever reaching the liver, preventing the needed chemicals from being absorbed.

Prescription drugs are not the only culprits here. Grapefruit juice, for example, inhibits cytochrome P450 3A4, the same enzyme that metabolizes estrogen and many statins prescribed to lower cholesterol, whereas the herbal supplement Saint-John's-wort boosts the activity of this enzyme. The result, in either case: unpredictable

Vocabulary

informative 有意義な
disparate 異種の, 共通点のない

metabolize 代謝する
trouble spots 問題の起こりやすい場所
intestine 腸

enzyme 酵素
cytochrome P450 シトクロム P450

estrogen エストロゲン
statin スタチン
Saint-John's-wort セント・ジョーンズ・ワート(セイヨウオトギリソウ)

variations in the potency of the medications.

Studies show that once more than four drugs are introduced to the body, the potential for adverse reactions increases exponentially. The trick to avoiding unwanted consequences from drug interactions, says Douglas S. Paauw, who teaches internal medicine at the University of Washington, is "knowing when you're stepping into dangerous territory."

BANKING ON DATA

And therein lies the rub. The U.S. Food and Drug Administration does maintain a record of reported drug side effects and possible interactions through its Adverse Event Reporting System. But the agency does not know of all, or even most, of the complications—or, indeed, whether or not the reported problems are merely chance events. Clinical trials of new drugs usually do not reveal any issues before a drug is approved, because they are relatively short, focus on a single medication and enroll a small number of participants. As a result, to learn of a possible new interaction, the FDA has to rely on prescribing physicians to take the time to announce problems to the Adverse Event Reporting System.

Nigam H. Shah, who teaches biomedical informatics at the Stanford University School of Medicine, hopes to improve the odds of discovery by collecting information about specific online searches performed by consumers on the Internet and by physicians on a pharmaceutical Web site called UpToDate. Using more than 16 million pieces of data—electronic records of diagnoses, prescriptions, clinical notes, and the like—on nearly three million people, Shah and his colleagues recently published a previously unsuspected association between heart attacks and a group of popular heartburn medications sold under such brand names as Prilosec and Prevacid; Shah's computer program calculated a 16 percent increase in heart

Vocabulary

exponentially 指数関数的に, 急激に

rub 困難, 厄介なこと
U.S. Food and Drug Administration 米食品医薬品局

clinical trial 臨床試験

enroll 被験者として登録する

heart attack 心臓発作
heartburn medication 胸焼けの薬

attacks with these types of drugs, which are prescribed more than 21 million times every year in the U.S. By definition, such a correlation does not prove causation, however, and the drug's label information has not been changed.

GENES AND A BOTTLE

Improved abilities for tapping the wealth of information available in human DNA may one day dramatically enhance the power to predict who will suffer most from drug interactions. "Everybody metabolizes drugs a little differently," Paauw says. And now advances in computational biology are beginning to link variations in our genes to differences in how our body absorbs, distributes, metabolizes and eliminates specific medications.

At Duke University's Center for Personalized and Precision Medicine, geneticist Susanne Haga is investigating how to use this quickly accruing genetic knowledge to improve safety, starting at the drugstore. In a recent unpublished survey, Haga found that 17 percent of responding pharmacists had offered or used results from genetic tests (which do not need to be prescribed by a physician) within the previous 12 months. For example, many pharmacists now offer such a test to patients filling prescriptions for clopidogrel, a blood thinner, to confirm the absence of gene variants that could interfere with the drug's action.

Haga is not trying to find previously unknown side effects. Instead she wants to make sure that known genetic complications are widely understood and identified as needed. To facilitate genetic testing and analysis by local pharmacists, Haga recently started the Community Pharmacist Pharmacogenetic Network. Still under development, the network's Web site—rxpgx.com—helps pharmacists access and interpret genetic tests. Although no national databank for collecting such information exists as

Vocabulary

correlation 相関関係
causation 因果関係

pharmacist 薬剤師

clopidogrel クロピドグレル（抗血栓薬の一種）

of yet, Haga hopes that the Precision Medicine Initiative, a project led by the National Institutes of Health, will help lay the foundation for such an effort.

Because prescriptions often come in batches and multiple genes can affect a single drug, Haga envisions a future in which we get tested in advance for key genetic variants that affect our body's ability to process different drugs. But these kinds of tests—which would allow individuals and their physicians to obtain the information whenever needed—are costly, which means that for now, one-off tests suited to elicit information about a single, specific prescription are the more viable option.

INTERIM STEPS

Meanwhile the FDA is trying other approaches for identifying potentially dangerous drug interactions before they occur. At the Center for Drug Evaluation and Research, deputy director Shiew-Mei Huang and others are creating computer models that use clinical research data to calculate how one drug will alter the concentration of another when both drugs are metabolized by the same enzyme. Armed with the concentration and the time it takes for the drugs to move through the body, mathematicians can predict how they will interact.

The approach is bearing fruit. The information sheet for the anticancer drug ibrutinib warns that its concentration could increase drastically if taken with erythromycin, a CYP3A inhibitor, and decrease with efavirenz, an HIV drug. These alerts were generated through computer calculations, not by clinical studies of the effects of both drugs taken simultaneously.

The trouble is that many companies and academics are likely to resist Huang's invitation to share the necessary data for a drug in development—such as its metabolic pathway or its most effective dose—to create useful com-

Vocabulary

one-off 1回限りの, その都度の
elicit 引き出す

ibrutinib イブルチニブ (抗がん剤の一種)
erythromycin エリスロマイシン
efavirenz エファビレンツ (抗HIV薬の一種)

metabolic pathway 代謝経路
dose 投与量, 用量

puter models. Such information could include proprietary data, and sharing it might give competitors an edge.

The FDA is also trying new ways to make the drug information packaged with prescription medicines, termed drug "labels," more useful to prescribers and the public. The aim is to give patients clearer warnings about possible drug interactions and easy-to-understand recommendations about how, for example, a dose should be altered (based on computer modeling) when a second drug is introduced. According to Huang, some of the new labeling changes are already being used for some recently approved drugs.

Vocabulary

gine ~ an edge 〜を有利にする

洗面台のキャビネットに洗面用品と一緒にエクセルの表をしまっておくのはちょっと奇妙かもしれないが，76歳のパインズ（Barbara Pines）にとってそれは，自分と夫が使っている処方薬と大衆薬，サプリメントを把握しておくための最適な方法だ。一覧表には20の薬の名が，その強度と服用頻度，目的とともに整理してある。「新しい先生にかかるときには必ずこれを印刷して持っていくの」と彼女はいう。

65歳以上の米国人の40％は5種を超える処方薬を服用しており，パインズはその一人だ。また，処方薬を飲んでいる人は高齢者だけではない。2014年に米国の薬局で調剤された処方箋は40億通を超える。国民1人あたり13件近い。

複数の薬を服用する場合，医師も患者も十分には気づいていない特別なリスクが生じる。ある組み合わせで医薬品（処方薬に限らない）を服用すると，単独では生じない副作用を引き起こすのだ。過去20年の研究から，こうした「薬物相互作用」が薬の副作用の30％以上の原因になっていることが示唆された。あいにく医薬品メーカーは，飲み合わせの悪い他の薬と一緒に新薬が服用されるケースを予想できるとは限らない。サプリメントや食品との組み合わせについてはなおさらだ。このため，予期せぬ死亡事例が出て初めて，飲み合わせの危険が判明する例もある。

副作用のすべてが致死的というわけではないものの，危険な組み合わせから人々を守る新たな取り組みが始まっている。共通点のない巨大な異種データから有意義なパターンを見つけ出すことがカギとなる。

薬とその代謝

薬物相互作用はふつう，人体が医薬品を分解（代謝）する際に生じる。問題が生じやすい場所は，消化した薬を血流に放出する小腸と，大半の薬を分解する肝臓だ。

肝臓での薬の分解は主に「シトクロムP450」と呼ばれる一群の酵素の作用による。「CYP（シップ）」と略称されるこのファミリーに属する酵素は約50種あるが，そのうちわずか6種類によって既知の医薬品の90％が分解される。だが，

3 潜むリスク

2つの薬が同じ酵素による処理を必要とすると，問題が生じうる。片方の薬がこの酵素の活性をブロックした場合には，他方の薬が分解されずに血中に過剰なまま残ることになる。逆に，片方の薬によって酵素の活性が高まると，他方は急速に分解されて効果が落ちるだろう。また，2種類の薬が小腸で結びつき，必要な成分が吸収されずに終わることもありうる。

問題を起こすのは処方薬だけではない。例えばグレープフルーツジュースはシトクロムP450 3A4という酵素を阻害する。女性ホルモン剤のエストロゲンやコレステロール低下薬スタチンを代謝する酵素だ。一方でハーブのセント・ジョーンズ・ワート（セイヨウオトギリソウ）を含むサプリメントはこの酵素の活性を高める。この結果，どちらのケースも薬の効き目に予測のつかない変化が生じる。

5種類以上の薬が体内に入ると副作用の危険が指数関数的に高まることが研究から示されている。ワシントン大学（シアトル）で内科学を教えているパァウ（Douglas S. Paauw）は，薬物相互作用の好ましくない効果を避けるには「危険地帯に踏み入った段階でそれを知る必要がある」という。

埋もれた危険を探り出す

そして，それが難しい。米食品医薬品局（FDA）は有害事象自発報告システム（AERS）を通じて薬の副作用と相互作用の報告を記録しているが，すべての事例を把握しているわけではない。報告されない例がかなりあるうえ，それらの事例（報告・未報告にかかわらず）が必然なのか単なる偶発事例なのかも不明だ。通常，認可前の臨床試験で新薬の問題が明らかになることはない。臨床試験が比較的短期間で，1つの薬に的を絞っており，被験者の数も少ないためだ。このためFDAが新たな薬物相互作用の可能性を知るには，薬を処方した医師がAERSに問題を報告するのを待つしかない。

スタンフォード大学医学部でバイオメディカル・インフォマティクスを教えているシャー（Nigam H. Shah）は発見の確率を高める方法として，一般消費者がインターネット上で行った特定の検索や，医師が「UpToDate」という医薬品サイトで行った検索に関する情報を収集するのが有効だろうと期待している。

シャーらは最近，300万人近い人々に関する1600万件を超える情報（診断，処方薬，臨床医による留意点などの電子データ）を用いて，プリロセックやプレバシドといった商品名で販売されている一般的な胸焼けの薬と心臓発作の間にこれまで予想されていなかった関連を見いだして論文発表した。シャーのコンピュータープログラムは，これらのタイプの薬（米国で毎年2100万件以上も処方されている）の服用者では心臓発作が16％多いことをはじきだした。ただし，こうした相関は因果関係を証明したものではもちろんなく，薬に表示される注意書きは以前と変わっていない。

ゲノムの知識を役立てる試み

ヒトのDNAから豊かな情報を引き出す技術が進展すると，薬物相互作用の起こりやすい人を予測できるようになるかもしれない。「薬の代謝は人それぞれで少し異なる」とパァウはいう。いまや計算生物学の進歩によって，遺伝子の変異と，人体による特定の薬剤の吸収・分配・代謝・分解を関連づける研究が始まっている。

デューク大学個別化・高精度医療センターの遺伝学者ヘイガ（Susanne Haga）は，これら急速に蓄積が進む遺伝学的知識を安全性向上に役立てる方法を研究している。注目したのは薬局だ。ヘイガは最近の未発表研究で，調査に回答した薬剤師の17％が過去12カ月間に遺伝子検査（医師に処方されたものには限らない）を実施するか検査結果を利用したことを突き止めた。例えば，多くの薬剤師は抗血栓薬クロピドグレルを調剤するにあたって，患者にこの薬の作用を阻害する遺伝子変異がないことを検査で確認している。

ヘイガは未知の副作用を見つけようとしているのではない。既知の遺伝的合併を周知して特定できるようにしたいと考えている。地域の薬局による遺伝子検査と解析を助けるため，「地域薬剤師ゲノム薬理学ネットワーク」というウェブサイト（rxpgx.com）を立ち上げた。サイトはまだ開発中だが，薬剤師が遺伝子検査を利用・解釈する手助けをする。この種の情報を集積した全国的なデータベースはまだないが，米国立衛生研究所（NIH）が主導している「高精度医療イニシアティブ」がその基盤となるだろうとヘイガは期待している。

3 潜むリスク

1通の処方箋は複数の薬を含むことが多いし，ある薬に影響しうる遺伝子は複数あるので，将来は様々な薬剤の代謝に影響する主な遺伝子変異を前もって検査しておくのが望ましいとヘイガはみる。だがこの種の検査は費用がかかるので，当面は特定の処方薬に関する情報を引き出す検査をその都度行うのが現実的だ。

当面の対処法

一方，FDAは危険な薬物相互作用を特定する別の方法を試している。FDA医薬品評価研究センターは副所長のフアン（Shiew-Mei Huang）らが臨床データを用いて，同じ酵素によって代謝される2種類の薬剤が互いに相手の濃度をどう変えるかを計算するコンピューターモデルを開発している。濃度の値と，これらの薬が体内に行き渡るのにかかる時間をもとに，どんな相互作用が生じるかを数理的に予測可能だ。

この方法は実際に実を結びつつある。抗がん剤イブルチニブの情報シートは現在，CYP3A阻害作用のあるエリスロマイシンと併用すると濃度が劇的に高まり，抗HIV薬のエファビレンツと併用すると濃度が低下すると警告している。これらはコンピューター計算によって判明したもので，薬を実際に併用した臨床研究の結果ではない。

このアプローチの問題は，多くの製薬企業と学術研究者が開発中の薬についてデータ（代謝経路や適切な用量など）を共有するのを拒む可能性が大きく，有用なコンピューターモデルができない懸念があることだ。こうした情報は企業秘密で，公開するとライバルを利することになる。

FDAはまた，処方薬に表示される薬理情報（いわゆる「ラベル」）を医師と一般人にとってより有意義なものに改良する方法も試している。起こりうる薬物相互作用を患者にはっきり警告し，2つ目の薬を追加した場合に以前からの薬の用量をどう変えるべきか（コンピューターモデルによって推定）をわかりやすく助言するのが目的だ。フアンによると，最近承認された新薬にはすでにそのようなラベルが使われている例がある。

When DNA Means "Do Not Ask"
遺伝子検査のジレンマ

包括的遺伝子検査の普及につれ,"予想外の発見"への対処が問題になってきた

D. F. マロン（SCIENTIFIC AMERICAN 編集部）

掲載：SCIENTIFIC AMERICAN January 2015, 日経サイエンス 2015 年 6 月号

Last spring Laura Murphy, then 28 years old, went to a doctor to find out if a harmless flap of skin she had always had on the back of her neck was caused by a genetic mutation. Once upon a time, maybe five years ago, physicians would have focused on just that one question. But today doctors tend to run tests that pick up mutations underlying a range of hereditary conditions. Murphy learned not only that a genetic defect was indeed responsible for the flap but also that she had another inherited genetic mutation.

This one predisposed her to long QT syndrome, a condition that dramatically increases the risk of sudden cardiac death. In people with the syndrome, anything that startles them—say, a scary movie or an alarm clock waking them from a deep slumber—might kill by causing the heart to beat completely erratically.

Doctors call this second, unexpected result an "incidental finding" because it emerged during a test primarily meant to look for something else. The finding was not accidental, because the laboratory was scouring certain genes for abnormalities, but it was unexpected.

Murphy, whose name was changed for this story, will most likely have plenty of company very soon. The

Vocabulary

mutation 変異

hereditary conditions 遺伝性疾患

predispose 〜に罹患しやすくする
long QT syndrome QT 延長症候群

slumber 眠り
erratically 常軌を逸して

incidental finding 偶発的所見

scour あれこれ探す

company 仲間, 同類

growing use of comprehensive genetic tests in clinics and hospitals practically guarantees an increasing number of incidental discoveries in coming years. Meanwhile the technical ability to find these mutations has rapidly outpaced scientists' understanding of how doctors and patients should respond to the surprise results.

UNKNOWN UNKNOWNS

Incidental findings from various medical tests have long bedeviled physicians and their patients. They appear in about a third of all CT scans, for example. A scan of the heart might reveal odd shadows in nearby lung tissue. Further investigation of the unexpected results—either through exploratory surgery or yet more tests—carries its own risks, not to mention triggering intense anxiety in the patient. Follow-up exams many times reveal that the shadow reflects nothing at all—just normal variation with no health consequences.

What makes incidental findings from genetic tests different, however, is their even greater level of uncertainty. Geneticists still do not know enough about how most mutations in the human genome affect the body to reliably recommend any treatments or other actions based simply on their existence. Furthermore, even if the potential effects are known, the mutation may require some input from the environment before it will cause its bad effects. Thus, the presence of the gene does not necessarily mean that it will do damage. Genetics is not destiny. In Murphy's case, her mutation means that she has a roughly 50 to 80 percent chance of developing long QT syndrome, and the presence of the mutation alone is not a sure indicator she will be afflicted, says her physician, Jim Evans, a genetics and medicine professor at the University of North Carolina School of Medicine. To be safe, he has advised her to meet with a cardiac specialist to talk about next steps, including possibly starting beta-blocker drugs to regularize her heart rate.

Vocabulary

comprehensive 包括的な

outpace しのぐ, 追い越す

bedevil 苦しめる

exploratory surgery 診査手術

uncertainty 不確実性

afflict 悩ます, 苦しめる

to be safe 大事を取って

beta-blocker ベータ遮断薬

遺伝子検査のジレンマ

The incidence of hard-to-interpret results is expected to rise because the cost of surveying large swaths of the genome has dropped so low—to around $1,000. It is typically less expensive to get preselected information about the 20,000 or so genes that make up a person's exome—the section of the genome that provides instructions for making proteins—than to perform a more precision-oriented test that targets a single gene. As a consequence, scientists and policy makers are now scrambling to set up guidelines for how much information from such testing to share with patients and for how best to help them deal with the inevitable incidental findings.

Before making any definitive recommendations, however, they need to know how often genetic results produce such findings. To that end, Evans is heading up the NCGENES clinical trial, part of a larger effort by three organizations, including the University of North Carolina School of Medicine. Of the roughly 300 patients who have received genetic information since Evans started ordering whole exome tests a couple of years ago, he says, six of them (or 2 percent) had incidental findings that required further testing or decisions about treatment.

Separately, Christine Eng, medical director of the DNA Diagnostic Laboratory at the Baylor College of Medicine, says her team has conducted more than 2,000 whole exome tests since October 2011 with about 95 incidental findings. "That's an incidence of about 5 percent," she notes. Most of the findings did not require immediate action. Usually they prompt more frequent screening tests,

Vocabulary

swath 大きな部分

exome
▶ Technical Terms

screening test スクリーニング検査, 健診

Technical Terms　エクソーム（**exome**）　ゲノムDNAのうち, タンパク質合成の指令をコードした配列（タンパク質に翻訳される配列）すべてを指す。遺伝子はそれぞれ特定のタンパク質を作るための情報を含んでいるが, mRNAを経てタンパク質に翻訳されるコード部分と, タンパク質のアミノ酸配列と直接には無関係で翻訳されない部分が複数に分かれて並んでいる。前者はエクソン（exon）, 後者はイントロン（intron）と呼ばれ, エクソームはエクソンの総体のことだ。-omeはgenome（ゲノム）やプロテオーム（proteome, ある生物のタンパク質の総体）などというように,「全体」「総体」を意味する。

often for breast cancer or colorectal cancer.

BALANCING ACT

In the hope of minimizing the number of people forced to cope with incidental findings, the American College of Medical Genetics and Genomics (ACMG) in 2013 proposed regularly returning results on 56 genes from comprehensive genetic tests. The professional group felt that there was enough—though by no means conclusive—information about these specific mutations to merit letting patients know if they had tested positive for them. In other words, the mutations "met a standard of relatively high likelihood of being disease-causing." The list included genetic variants that have been strongly linked to retinoblastoma (cancer of the eye), hereditary breast cancer and long QT syndrome. The ACMG believed that its guidance would give physicians a shortcut so they would not need to haphazardly guess which mutations had a strong enough link to a given malady to tell patients about the results.

Such advice is particularly important given how often children undergo genetic tests nowadays. "About 80 percent of our cases are pediatric-aged, so the incidental findings are being found in the children, and many of the conditions are adult-onset conditions," Eng says. Families given such information about their children then may have to wait decades before they can do anything about it or decide when, if ever, to start considering treatment for a disorder that may not ever develop.

Yet a year after issuing its guidance, the ACMG produced an addendum: patients should have the opportunity to opt out of having information about even that short list of analyzed genes. "When families are given a choice, a very large percentage of them want this information, but there are some individuals who feel they do not want this information, so I think this option is a good

Vocabulary

colorectal cancer 大腸がん

American College of Medical Genetics and Genomics 米国臨床遺伝学会

by no means 決して〜ない

retinoblastoma 網膜芽細胞腫（目のがん）

haphazardly 無定見に
malady 疾患

pediatric 小児科の

onset 襲来, 発症

addendum 追加, 補遺
opt out 不参加を選ぶ, しないことに決める

one," says Eng, who was not on that decision-making board.

For her part, Murphy is still grappling with how to respond to her incidental finding. She is not yet 30, and she finds it hard to imagine being young and carefree and on beta blockers. "Generally, I'm a very healthy person. I was doing just fine until now, so why does it matter that I found this out?" she asks. "I've been giving it a lot of thought, and if I hadn't gotten [the test] done, I might never have known about this. Now I'm wondering if I really want a lifestyle change. It's a lot to think about." Yet the hope is that Murphy's experience, and those of other patients, will help geneticists decide which tests to include in future gene scans and better prepare patients and health care workers for dealing with any unwelcome surprises.

Vocabulary

grapple with 取り組む

carefree 屈託のない

昨年春，28歳のローラ・マーフィー（仮名）は病院に医師を訪ね，ずっと前から首の後ろについている無害な皮膚のフラップが遺伝子変異によるものかどうかを調べてもらった。かつて，5年ほど前までなら，医師はその点だけを調べただろう。だが現在では，様々な遺伝性疾患の背景となる変異を探し出す検査を行うことが多い。マーフィーは検査の結果，皮膚フラップがまさに遺伝子の欠陥によるだけでなく，自分が別の変異を受け継いでいることを知った。

この変異があると「QT延長症候群」という病気になりやすい。突然の心臓発作で死亡するリスクが著しく高まる病気だ。怖い映画を観たり，目覚まし時計のけたたましい音で起こされたりしてびっくりすると，心臓が滅茶苦茶に拍動して命を落としかねない。

医師たちはこうした予想外の副次的検査結果を「偶発的所見」と呼んでいる。本来とは別の目的でなされた検査で偶然に見つかったからだが，検査自体は遺伝子の異常をあれこれ探し回っていたわけだから，発見そのものは偶然ではなく必然だ。ただ予想外だっただけである。

3 潜むリスク

間もなくマーフィーと同じような経験をする人が急増するだろう。包括的遺伝子検査を利用する診療所や病院が増えており，偶発的所見の件数が増えるのは確実だ。その一方で，医師と患者がそうした驚きの検査結果にどう対処すべきかについての科学的理解は検査技術の急進展にまったく追いついていない。

避けがたい不確実性

遺伝子検査に限らず，様々な検査で生じる偶発的所見が以前から医師と患者を悩ませてきた。例えば CT 検査の 1/3 で偶発的所見が生じる。心臓を撮像した際に近くの肺組織に奇妙な影が見つかるといった例で，この予想外の結果をさらに調べる処置は，診査手術であれ精密検査であれ，それ自体がリスクを伴う。患者本人が強い不安を感じるのは言うまでもない。そして追加検査の結果，影は画像によく生じる変動にすぎず健康にはまったく影響なしと判明することが多い。

ただ，遺伝子検査で生じる偶発的所見は不確実性がより大きい。ヒトゲノムに見られる変異の大半については，それが人体にどう影響するのかが十分にわかっておらず，変異があるというだけで治療などの介入を推奨できる状況にはない。さらに，変異の影響が知られている場合も，実際に悪影響が生じるには変異が環境から何らかの刺激を受ける必要があるかもしれない。したがって，変異遺伝子の存在はそれが害を及ぼすことを必ずしも意味しない。遺伝子は運命ではない。マーフィーの場合，QT 延長症候群の発症確率はざっと 50%から 80%になるとみられるが，変異の存在だけで必ず発症することにはならないとノースカロライナ大学医学部で遺伝学・医学の教授を務める医師のエバンズ（Jim Evans）はいう。彼はマーフィーに，念のため心臓の専門医に診てもらって心拍を安定化するベータ遮断薬を服用するなどの対処法を検討するよう助言した。

ゲノム解析のコストが大幅に下がったので，解釈の難しい結果に直面する例はさらに増えるとみられる。いまや 1000 ドルほどでゲノムの大部分を調べられる。エクソーム（タンパク質生産の指令を含むゲノム領域）を構成している 2 万種あまりの遺伝子に対象を絞って調べる検査は，何か 1 つの遺伝子を高精度で探し出す検査よりも，一般に安く実行できる。このため科学者や政府当局者は，そうした検査結果をどこまで本人に知らせるべきか，そして避けられない偶発的所見に本人が対処するのを助ける最善の方法は何かについて，指針作りを急いでいる。

しかし，その前にまず，どんな頻度でそうした検査結果が出るのかを知る必要がある。このためエバンズは，ノースカロライナ大学医学部など3機関による取り組みの一環として「NCGENES」という臨床試験を率いている。2年前にスタートして以降，全エクソーム検査を受けた約300人の患者のうち6人（つまり2%）に偶発的所見が生じ，追加検査や治療の決定が必要になったという。

　これとは別に，ベイラー医科大学DNA診断研究所の所長エン（Christine Eng）のチームが2011年10月以降に2000人以上に全エクソーム検査を行った結果，約95件の偶発的所見が生じた。「約5%ということだ」とエンは指摘する。ただし，ほとんどはすぐに対応する必要のないもので，乳がんや大腸がんに関する健診をより頻繁に受診するよう勧めるにとどまった。

求められるバランス技

　偶発的所見への対処に迫られる人をなるべく少なくしようと，米国臨床遺伝学会（ACMG）は2013年，包括的遺伝子検査の結果のうち本人にフィードバックする情報を56の遺伝子に限るよう提案した。検査で陽性と出てそれを患者本人に知らせることが有益な変異はこれで十分であるとの判断だ。言い換えれば，これらの変異は「病気を引き起こす可能性が標準以上に高い」。リストに挙がった変異は，網膜芽細胞腫（目のがん）や遺伝性の乳がん，QT延長症候群に強く関連づけられているものなどだ。このガイドラインが内科医の手引きとなり，どの変異が疾患に強く関連していて患者に結果を知らせる必要があるのか悩まずにすむだろうと米国臨床遺伝学会はみている。

　いまや多くの子供たちが遺伝子検査を受けるようになったことを考えると，こうしたアドバイスは特に重要だ。「約80%は小児科の対象年齢の患者なので，予想外の検査結果が子供たちに見つかっており，そしてその多くが大人になってから発症する病気だ」とエンはいう。そうした検査結果を知らされた子の家族は，何も対処することができないまま何年も不安にさらされ，病気に対する治療をいつ始めるかを思い悩み，結局のところその病気は発症しないかもしれない。

　米国臨床遺伝学会はガイドラインの発表から1年後に条項を追加した。リストに挙げられた少数の遺伝子の検査結果について，知りたくないと思う患

者は情報を知らされないという選択ができるようにすべきである，というものだ。「大半の家族は結果を知ろうとするだろうが，なかには知る必要はないと思う人もいるだろうから，この選択肢を設けるのはよいことだ」とエンはいう（エンはこの決定をした委員会には加わっていない）。

　マーフィーは依然として偶発的所見にどう対処すべきか悩んでいる。彼女は20代であり，ベータ遮断薬を服用しながら若くて快活な自分というものを想像しにくいと感じている。「総じて自分はとても健康。これまで元気にやってきたわけで，この検査結果を知ったからといって何だというの？」と問う。「この件についてはずいぶん考えた。もし検査を受けなかったら，これを知ることも決してなかったろう。本当にライフスタイルを変えたほうがよいのか，迷っている。考えることがたくさんある」。だが希望は，マーフィーや他の患者の経験が，将来の遺伝子検査でどの遺伝子を対象にしたらよいのかを専門家たちが決める助けとなり，患者と医療関係者がありがたくない驚きにうまく対処するのに役立つだろうということだ。

先端医療

Busting Blood Clots
危険な血栓を取り除く

Cancer Gene Tests Provide Few Answers
がん遺伝子検査のいま

A Surprising Fix for Sickle Cell
鎌状赤血球症に驚きの治療法

The Paradox of Precision Medicine
個別化医療の矛盾

The Not So Silent Epidemic
広がる睡眠時無呼吸症

Busting Blood Clots
危険な血栓を取り除く

血栓溶解薬には限界もある。除去・抑制の新手段が登場してきた

D. ヌーナン（サイエンスライター）

掲載：SCIENTIFIC AMERICAN October 2016, 日経サイエンス 2017 年 7 月号

Twenty years ago stroke doctors celebrated the arrival of a powerful new weapon: the clot-clearing drug tPA. It was hailed as a lifesaver and has proved to be one for hundreds of thousands of patients since. TPA was the first and is still the only medicine approved by the U.S. Food and Drug Administration for treating strokes caused by clots that block blood flow to the brain. But like so many medical marvels, tPA (which stands for tissue plasminogen activator) has turned out to have serious limitations. It needs to be administered within three hours of symptom onset, does not last long in the body before it loses effectiveness, can cause uncontrolled bleeding and often fails to break up large clots.

For many of the nearly 800,000 Americans who every year suffer ischemic strokes, as the brain blockages are called, these shortcomings can be deadly. Nearly 130,000 die. Sadly, there have been no good alternatives to tPA since it debuted.

Recently doctors and scientists have broken this long-standing clinical stalemate with new tools to put a

Vocabulary

stroke 脳卒中
tPA
▶ Technical Terms
hail 〜と認める
U.S. Food and Drug Administration 米食品医薬品局
clot 血栓
marvel 驚嘆すべきもの

administer 投与する
onset 病気の始まり

bleeding 出血

ischemic stroke 虚血性脳卒中

stalemate 手詰まり, 膠着状態

Technical Terms　組織プラスミノーゲン活性化因子（**tPA**）　タンパク質分解酵素プラスミン（その前駆体がプラスミノーゲン）を活性化することで, 血液の凝固に関与しているフィブリンというタンパク質を分解する物質。血栓溶解薬として使われている。

dent in those grim numbers. One innovation, a tiny wire device called a stent retriever, can be snaked up into the blood vessels leading to the brain to pull out large clots. "It's the first proven, effective treatment for acute stroke in a generation," says Jeffrey Saver, director of the Stroke Center at the University of California, Los Angeles. Approved by the FDA in 2012, the stent retriever got a boost this year when the journal *Stroke* reported data showing many more patients treated with a retriever resumed normal life than did patients who received tPA. (The retriever manufacturer, Medtronic, provided support for the studies. Neurologist Bruce Campbell of the Royal Melbourne Hospital in Australia, who co-led the analysis, notes that *Stroke* has "strict independent-peer-review processes" to guard against bias.) Researchers are also developing better clot-detection scans, as well as a technique involving magnetism that guides tPA directly to the problem. This method could help eliminate dangerous obstructions elsewhere in the body, as well as in the brain.

BIG CLOTS, BIG TROUBLE

Of all of tPA's drawbacks, the most troublesome is its inadequacy against big clots, which can block large blood vessels at the base of the brain; they cause about 25 to 30 percent of all strokes. Although it works well against smaller clots in narrower vessels, a safe dose of the drug—which is delivered intravenously—often does not last long enough in the bloodstream to dissolve the big clots, and increasing the dose raises the risk of bleeding. "All you need to see is one intracranial bleed from tPA, and you realize you've got to pause before you give that medication," says Thomas Maldonado, a clot specialist at New York University's Langone Medical Center.

This is where the stent retriever has an advantage. It is a narrow tube that can be threaded up from the femoral artery in the leg to the site of the clot. Then wire mesh on the end of the retriever, which expands like an accordion,

Vocabulary

put a dent in 減らす
grim ぞっとするような
stent retriever ステント型血栓回収デバイス

boost 後押し, 推奨, 増加

peer-review ピアレビュー, 同分野の専門家による評価

eliminate なくす
obstruction 塞栓

drawback 欠点
inadequacy 不適当, 不十分

dose 投与量, 用量
intravenously 静脈注射で

intracranial 頭蓋内の

femoral artery 大腿動脈

is pushed into the clot. The mesh tendrils keep the clot from breaking apart in the brain—which could be deadly—and help separate it from blood vessel walls. The device is next pulled out of the body, and the clot comes with it. (In years past, doctors had tried a device with a corkscrew tip but found it was not as good at clearing the clot.)

Another advantage the device has over the drug is that the time window for the use of the stent retriever after symptoms arise is double that of tPA—six hours instead of three. The *Stroke* analysis found that blood flow in a vessel blocked by a large clot was successfully restored in 236 of 306 patients, or 77 percent, treated with the stent retriever. With tPA alone, the success rate was around 37 percent.

Like all surgical interventions, the stent retriever carries the risk of complications. The main one is bleeding, which is why patients with high blood pressure and the strained vessels that go with it may not be candidates for the procedure. "There's also a chance of the guide wire or some other manipulation of the device poking through the blood vessel during the procedure," Saver says.

A much less common complication, Saver adds, is a piece of the clot breaking off as it is being pulled out, escaping into a new artery and causing a new stroke in a different area than the initial one. It happens in about 2 to 3 percent of cases, he says.

HELP FROM IMAGING

The damage that blood clots do is not limited to strokes. Every year as many as 900,000 people in the U.S. develop blood clots in their legs, called deep vein thrombosis (DVT). Aside from the localized discomfort and pain it causes, DVT can travel to the lungs and become potentially lethal pulmonary embolisms, which kill

Vocabulary
tendril 巻きひげ状のもの

restore 回復する

intervention 介入処置
complication 合併症

poke through 突き破る

artery 動脈

deep vein thrombosis 深部静脈血栓症
pulmonary embolisms 肺血栓塞栓症

an estimated 100,000 people annually. These two types of clots are usually treated with the anticoagulants heparin (for acute situations) and warfarin (for long-term problems), and surgery may be used in serious cases. The FDA approved tPA for the acute treatment of lung clots in 2002; although it carries the usual risks and complications, it can reduce the size of clots, which the anticoagulants cannot do. The drug is also gaining traction as a treatment for some cases of leg blockages.

Vocabulary
anticoagulant 抗凝固薬
heparin ヘパリン
warfarin ワルファリン

traction 魅力, 影響力

Knowing more precisely where these clots are would help doctors go after them: location can affect choice of drugs or other treatments. Alas, current imaging methods have limitations. Existing technologies are "very good if we know where we're looking," says Peter Caravan, a radiologist and co-director of the Institute for Innovation in Imaging at Massachusetts General Hospital, but there is currently no single whole-body test that can spot blood clots anywhere that they might form. Ultrasound is the first choice for finding a clot in the legs, and computed tomography (CT) scans readily detect pulmonary embolisms. CT is also the primary imaging choice for patients who arrive at the hospital with symptoms of stroke. "But if we don't know where to look, we have to subject the patient to a battery of tests," Caravan says, a costly and time-consuming process that can delay critical treatment.

go after 追跡する

radiologist 放射線科医

ultrasound 超音波（画像法）
first choice 第一選択
computed tomography コンピューター断層撮影法

a battery of 一連の

To address the problem, Caravan and his team have invented an imaging agent that, when injected into the bloodstream, binds to fibrin, the meshlike protein that forms clots, and makes it visible to a scanner. It has potential applications for all clots, including those that cause strokes. "About a third of ischemic strokes are of unknown origin," Caravan says. "You may think, at first, 'So what? You had the stroke—why do you care where it came from?' But it's really about preventing that second stroke. Your biggest risk of having a stroke is if you've already had one."

imaging agent 造影剤
fibrin フィブリン

Because the experimental probe binds to fibrin (and "lights up" in a positron-emission tomography scan), it can help establish how dangerous a clot is: younger clots, which have more fibrin than older ones, are less stable and more likely to travel to the lungs. They can also make it up to the brain, triggering a stroke. Further, tPA is more effective against fibrin-rich young clots than it is against older clots, and so the probe could help determine which clots to attack with the drug. After a series of animal experiments, researchers began safety testing of the new agent in healthy human subjects this past spring.

Some doctors think that tPA could work faster and prevent strokes more successfully if the drug could be guided swiftly and efficiently to the clot rather than simply being injected into the bloodstream. Researchers at Houston Methodist Hospital are experimenting with a way to transport tPA to the clot while protecting it from the body's defenses, which degrade the drug. They are experimenting with iron oxide nanoparticles, stuffed with tPA and "biochemically camouflaged" with a coating of the naturally occurring blood protein albumin. The albumin jacket fools the body's defenses and gives the tPA extra time to work on the clot; the iron oxide enables monitoring with magnetic resonance imaging, remote guidance of the nanoparticles with external magnetic fields and magnetic heating at the site to accelerate clot dissolution. And because the tPA does not degrade while it is being ferried to the clot inside the iron oxide, the dose can be smaller, reducing the risk of hemorrhage. Results with human tissue cultures and animal models have been promising, and clinical trials in humans are planned.

Of course, stopping blood clots from forming in the first place would be even better. There is a growing list of clotting conditions caused by genetic mutations, and researchers around the country, including a team at N.Y.U. Langone, are analyzing the role that genes play in

Vocabulary

probe プローブ, 造影剤
positron-emission tomography 陽電子放射断層撮影装置

iron oxide 酸化鉄
nanoparticle ナノ粒子

albumin アルブミン

magnetic resonance imaging 磁気共鳴画像法

dissolution 溶解, 分解
degrade 劣化する

hemorrhage 出血

genetic mutation 遺伝子変異

clots. The goal is to develop a genetic test that would show if a person is at increased risk and to offer preventive treatment such as an anticoagulant. This approach could hinder the blockages, making elaborate feats with wires and magnets unnecessary.

Vocabulary

preventive 予防的な
hinder 妨げる
elaborate 手の込んだ

20年前，脳卒中の専門医たちは強力な新薬の登場を祝った。血栓溶解薬の組織プラスミノーゲン活性化因子（tPA）だ。決定的な救命薬とされ，実際これまでに数十万人の命を救った。血栓が脳の血流を妨げる脳梗塞の治療薬として米食品医薬品局（FDA）が初めて認可した薬で，現在でも唯一だ。だが多くの画期的医薬と同様，tPAにも重大な制約が判明した。発作から3時間以内に投与する必要があるほか，体内で有効性を割に早く失い，抑えられない出血を招く場合がある一方で大きな血栓は分解できないことが多い。

これらの欠点は，米国で年間80万人近くに上る脳梗塞（虚血性脳卒中とも呼ばれる）の患者の多くにとって致命的となる。同13万人近くが命を落としており，残念ながらtPAに代わる有効な薬はない。

長年のこの手詰まりを打開し，死亡者を減らす新手段が近年にいくつか開発された。一例は「ステント型血栓回収デバイス」という小さなワイヤ装置で，脳に至る血管に挿入して，大きな血栓を体外に引き出すことができる。「急性の脳卒中に有効性が立証された過去30年で初めての新治療法だ」とカリフォルニア大学ロサンゼルス校・脳卒中センター所長のセイバー（Jeffrey Saver）はいう。2012年にFDAの認可を得たこのデバイスの治療成績は上々で，tPAを投与した場合に比べ，通常の生活を送れるまでに回復した人がずっと多いことが2016年の*Stroke*誌に報告された〔この一連の研究は装置の製造元であるメドトロニック社が支援しているが，論文共著者でオーストラリアにあるロイヤルメルボルン病院で神経科医を務めているキャンベル（Bruce Campbell）によると，*Stroke*誌はバイアスを排するため「第三者による厳格なピアレビュー」を行った〕。このほか，血栓を検出する新たな画像技術や，磁気を利用してtPAを患部に直接導く方法の

大きな血栓，大問題

tPAの最大の欠点は，脳の基底部にある太い血管を詰まらせるような大きな血栓に対する効果が不十分なことだ。脳梗塞の20〜30%はこうした大きな血栓が原因となっている。細い血管内の小さな血栓にはうまく働くものの，安全な用量（投与は静脈注射による）では血中に長くは存在しないことが多く，大きな血栓は溶解できない。用量を増やすと出血の危険が高まる。「tPAが頭蓋内出血を招いていないか監視して，気づいたら投与を中断する必要がある」とニューヨーク大学ランゴン医療センターの血栓の専門家マルドナド（Thomas Maldonado）はいう。

そこで頼りになるのがステント型血栓回収デバイスだ。脚の大腿動脈から挿入して血栓のある部分まで送り込める細いチューブで，患部で先端のワイヤメッシュがアコーディオンのように広がって血栓をとらえる。メッシュが血栓に巻き付いて脳内で分解するのを防ぎ（細かく分裂すると致命的な結果につながる），血管壁から引きはがす。その後デバイスを体外に引き出すと，血栓が一緒にくっついて除去される（かつてコルクスクリュー型の装置も試されたが，血栓をあまりうまく除去できなかった）。

発作から3時間以内に投与する必要があるtPAに対し，ステント型血栓回収デバイスは6時間以内でよく，時間的余裕が大きいのも利点だ。*Stroke*誌に掲載された先の解析によると，大きな血栓によって妨げられていた血流が306人の患者中236人，つまり77%で適切に回復した。tPAだけの場合，この率は約37%だった。

ただ，他の外科的な処置と同じくリスクも伴う。主な問題は出血で，高血圧によって血管に負荷がかかっている患者には適用できない。「ガイドワイヤなどデバイスの操作部が血管壁を突き破る恐れもある」とセイバーはいう。

た，まれではあるが，デバイスによって引き出されている最中に血栓が崩れ，かけらが別の動脈に入り込んで新たな場所に脳梗塞を生じることがある。処置例の2〜3%でこれが起こるという。

画像の助け

血栓が引き起こすのは脳梗塞だけではない。米国では毎年90万人もの人が脚に「深部静脈血栓症」という塞栓を生じる。この障害は局部の痛みと不快感に加え，血栓のかけらが肺に移動して「肺血栓塞栓症」という致命的な塞栓を起こす場合がある。肺血栓塞栓症による米国の死者数は年間10万人と推定されている。これら2種類の塞栓症の治療には通常，抗凝固薬の「ヘパリン」と「ワルファリン」が使われ（ヘパリンは急性症状向け，ワルファリンは慢性向け），重篤な場合には外科手術を行う。FDAは2002年，急性の肺血栓塞栓症の治療にtPAを認可した。他の症例に用いる場合と同様のリスクはあるものの，血栓を縮小できる（これは抗凝固薬では不可能）。脚部に生じた一部の血栓塞栓症についても有効だろうと期待を集めている。

これらの血栓の位置が正確にわかれば，その後の追跡に役立つだろう。患部の位置によって薬や処置の選択が変わりうるのだが，残念ながら現在の画像技術には限界がある。既存技術は「どこを調べればよいかわかっている場合には十分」なのだが，どこかに潜んでいるかもしれない未知の血栓を特定できる単独の全身検査技術はないのだと，マサチューセッツ総合病院で医療画像革新研究所の共同所長を務めている放射線科医キャラバン（Peter Caravan）はいう。　脚部の血栓を見つけるには超音波画像が第一選択で，肺の血栓はコンピューター断層撮影装置（CT）ですぐに検出できる。脳卒中で病院に担ぎ込まれた患者についてもCTが使われる。「だが，調べるべき場所が不明な場合は一連の検査を繰り返さねばならない」とキャラバンは説明する。費用だけでなく時間もかかり，重要な治療が遅れる恐れがある。

この問題に対処するため，キャラバンらは新たな造影剤を開発した。血液中に注入すると血栓を構成しているフィブリンという網目状のタンパク質と結合し，フィブリンがスキャナーで撮像できるようになる。脳卒中を引き起こすものを含めすべての血栓に使える可能性がある。「虚血性脳卒中のおよそ1/3は血

栓の出所がわからない」とキャラバンはいう。「『それがどうした？　すでに発作が起こってしまったのだから，出所なんかどうでもいいだろう』と思うかもしれないが，そうではない。次の発作を防ぐことが重要なのだ。脳卒中を起こした人は2度目を起こすリスクが大きい」。

この実験的な造影剤はフィブリンに結合して陽電子放射断層撮影装置（PET）で見ると"明るく光る"ため，その血栓がどれだけ危険かを判断するのに役立つ。形成間もない若い血栓は古いものに比べてフィブリンが多く，より不安定で肺に移動する可能性が高い。脳に行って卒中を起こす恐れもある。さらに，tPAは古い血栓よりもフィブリンが豊富な若い血栓に対して有効なため，この造影剤はtPAを使うのに適した血栓を見分けるのにも役立つ。一連の動物実験を経て，健康な人で安全性を評価する臨床試験が2016年春に始まった。

tPAを単に静脈注射するのではなく血栓の場所へ効率的に誘導すれば，短時間で効果が上がり脳卒中をうまく防げるだろう。ヒューストン・メソジスト病院の研究チームは，tPAを人体の免疫系（tPAを分解する）から守りながら血栓へ輸送する方法を実験している。酸化鉄のナノ粒子にtPAを詰め，血液中に見られるアルブミンというタンパク質でこれを覆って"生化学的にカムフラージュ"したものだ。アルブミンの上着が免疫系をだまし，tPAが血栓に作用する時間を稼ぐ。また酸化鉄は磁気共鳴画像装置でモニターできるうえ，外部磁場によってナノ粒子の行き先を遠隔操作でき，患部に到達したら磁気によって誘導加熱して血栓の分解を加速できる。さらにtPAは酸化鉄粒子内部で劣化を避けながら血栓に達するから，用量が少なくてすみ，出血のリスクを減らせる。ヒト培養組織とモデル動物での実験結果は有望で，臨床試験が計画されている。

もちろん，血栓の形成を最初から阻止できればさらによい。血栓症を引き起こす遺伝子変異がすでに数多く見つかっており，ニューヨーク大学ランゴン医療センターを含め全米の研究者がそれら遺伝子の働きを解析中だ。血栓が生じるリスクを判定し，抗凝固薬などの予防処置をすべきかどうかを判定できる遺伝子検査の開発が目標だ。これによって塞栓を防ぐことができたら，ワイヤや磁石を用いた精巧な治療術は不要になる。

Cancer Gene Tests Provide Few Answers
がん遺伝子検査のいま

進歩はしているが，まだまだの段階だ

J. ワプナー（サイエンスライター）

掲載：SCIENTIFIC AMERICAN September 2016, 日経サイエンス 2017 年 6 月号

Genetic tests for cancer have come a long way since they first entered the clinic in 1995. Back then, mutations in two genes—known as *BRCA1* and *BRCA2*—hinted at the crucial role that genetics can play in treatment decisions. Women carrying one of those mutations (and having a family history of breast or ovarian cancer) were much more likely than the general population to develop tumors in their breasts or ovaries. Then, as now, some of these women opted to have their breasts and ovaries removed before any malignant growths could arise.

In the intervening decades, researchers have come to recognize that most cancers are driven largely by abnormalities in genes. Genetic analysis of tumors has, therefore, become standard practice for many malignancies—such as breast, lung and colon cancer—because the information may help guide therapy. Clinicians have amassed a modest arsenal of drugs able to counteract some of the most common mutations.

Yet many patients learn that their cancers have mutations for which no drug exists. In fact, the roles many of these genetic changes play in cancer growth are poorly understood. Complex analyses of DNA done across a

Vocabulary

genetic test 遺伝子検査
mutation 変異

family history 家族歴
breast cancer 乳がん
ovarian cancer 卵巣がん
tumor 腫瘍

opt 選ぶ，〜すると決める

abnormality 異常

malignancy 悪性腫瘍
colon cancer 大腸がん
amass 蓄積する
arsenal 兵力，戦いのための備え
counteract 妨げる

range of cancer types have revealed a landscape rife with genetic mutations, and very little of this encyclopedic information is helping doctors to make treatment decisions. To date, the U.S. Food and Drug Administration has approved just 29 tests for specific mutations that can directly influence therapy.

Several major research collaborations are now making heroic efforts to identify more mutations that can serve as drug targets and to collect the information that will allow doctors to match many more patients with such targeted therapies. And earlier this year President Barack Obama announced the National Cancer Moonshot, a $1-billion initiative that includes funding for such efforts. The task is so large and complicated, however, that the gap between genetic knowledge and patient benefit is likely to widen for some time before the promised revolution in care becomes a reality for most people afflicted by cancer. "We're in a transition period," says Stephen Chanock, who directs the Division of Cancer Epidemiology and Genetics at the National Cancer Institute.

DRIVERS VS. PASSENGERS

The genetic changes that eventually trigger cancerous growth fall into two main groups. First, there are hereditary germ-line mutations, which people inherit from their parents. Second, there are somatic mutations, which arise over the course of one's life as a result of advancing age, cigarette smoking or other environmental influences. Although hereditary changes in DNA often lead to aggressive tumors, including some childhood cancers, these kinds of germ-line mutations are relatively uncommon. The vast majority of human cancers arise from somatic mutations.

Most somatic mutations turn out to be harmless; many are even repaired by the body's own quality-control processes. But some manage to wreak havoc, causing cells to reproduce uncontrollably. Many genes code for proteins, which do much of the work in cells. In the case of cancer, the harmful mutations tend to result in proteins that either actively promote excessive replication or fail in their usual job of putting the brakes on cell proliferation.

Researchers refer to the abnormal changes that are integral to a tumor's growth and survival as driver mutations; the others are known as passenger mutations because they appear to be unimportant and seemingly are just along for the ride. No one knows how many driver mutations are needed to promote each of the different kinds of cancers. One study determined that the average tumor requires as few as two or as many as eight driver mutations, whereas other studies found that tumors may frequently contain as many as 20 driver mutations.

EARLY SUCCESSES

Despite the difficulties of figuring out which genetic mutations are important in a given tumor, researchers began making progress in targeting specific cancer mutations by the late 1990s. Among the first such treatments were imatinib mesylate (brand name Gleevec), which undermines a common driver of chronic myeloid leukemia, and trastuzumab (brand name Herceptin), which addresses the *HER2* mutation responsible for about a quarter of breast cancers. Other customized therapies soon followed.

For the past three years patients with lung cancer have routinely been tested for an abnormality in a gene known as *ALK*. In as many as 7 percent of such patients, a genetic mistake that melds the *ALK* gene with another gene yields an abnormal protein that drives the tumor's growth. Drugs that block this mutant protein typically do a better job than standard chemotherapy at slowing the disease. Patients with normal *ALK* genes in their tumors

> **Vocabulary**
>
> reproduce 分裂・増殖する
> code for 〜をコードする
> replication 分裂・増殖
>
> proliferation 増殖
>
> integral to 〜に不可欠な
> driver mutation ドライバー変異
> passenger mutation パッセンジャー変異
> along for the ride 尻馬に乗って
>
> imatinib mesylate メシル酸イマチニブ
> undermine 阻害する
> myeloid leukemia 骨髄性白血病
> trastuzumab トラスツズマブ
>
> meld 融合する
> yield 生む, もたらす
>
> chemotherapy 化学療法, 抗がん剤療法

169

do not benefit from anti-*ALK* drugs at all.

Routine genetic tests have also helped people with melanoma, a form of skin cancer. About half of patients with melanoma have a mutation in the *BRAF* gene, which plays a role in the spread of cancer from the tumor to other parts of the body. In 2011 the FDA approved the first drug that inhibits the mutant BRAF protein. A recent study found that nearly 80 patients with metastatic melanoma who responded to the new treatment lived for an average of two years, much longer than the 5.3 months typically seen in such patients whose skin cancer has spread.

Sometimes a particular mutation allows doctors to steer clear of prescribing certain drugs. For example, colorectal cancers with mutations in the *KRAS* or *NRAS* gene typically do not respond to particular medicines because these genetic changes render those agents ineffective.

But there are several obstacles to further progress. Finding a genetic abnormality in a cancer is not enough—the aberration must be integral to the cancer's growth and survival. A reliable test for the mutation and a treatment that can exploit the mutation must exist. These requirements, it turns out, are a very tall order. Beyond the difficulty of figuring out which mutations drive the cancer, researchers also need to know which mutations tend to act later on. As a tumor grows, new mutations may appear. Each crop of abnormalities means separating the drivers from the passengers all over again, so that if one drug stops working, a subsequent genetic test can steer physicians to the next option.

Creating drugs to block driver mutations, likewise, is no small feat. Many abnormal proteins encoded by somatic mutations sit on the surface of cancer cells, within

Vocabulary

melanoma 黒色腫

inhibit 阻害する

metastatic 転移性の

steer clear of ～を避ける

colorectal cancer 結腸直腸がん

aberration 異常
integral to ～に不可欠

exploit 利用する

tall order 難しい注文

crop 群れ, グループ

feat 偉業, 妙技

easy reach of drugs. But others are buried deep within cells, and compounds small enough to slip inside a cell are typically too small to stick to their target proteins. This conundrum has left the most common driver mutations, such as *p53*, *RAS* and *MYC*, impossible to combat.

And the drugs that do successfully target somatic mutations have often led to meager extensions in survival time. If a single drug targeted to a specific driver mutation manages to shrink a tumor but leaves even one cell resistant to the drug behind, that cell can proliferate and create additional tumors unresponsive to the medicine. It may be, then, that certain cancers, as is true of HIV, will need to be treated with multiple drugs. Yet each drug that is added will come with its own costs and potential side effects. Researchers will need to figure out the optimal strategies.

The rarity of many somatic mutations also slows the transition from the laboratory to the clinic. Some mutations occur in less than 1 percent of patients with a certain type of cancer. Evaluating whether a drug could possibly address that mutation requires a clinical trial, but finding enough patients willing and able to enroll in such a study can take a long time.

NEW DIRECTIONS

All these challenges are spurring research methods, drug designs and infrastructure meant to hasten the expansion of precision genetic medicine. The approaches are also taking into account a new realization. Traditionally cancer has been defined by the location of where it first arose in the body—for example, in the breast or lung. But it turns out that mutations known to drive a particular type of malignancy in one part of the body are sometimes involved in cancers typically found elsewhere in the body.

Vocabulary

stick to 結合する

conundrum 難題

meager わずかな
survival time 生存期間

resistant 抵抗性の
unresponsive 薬に反応しない，薬の効かない

side effects 副作用

clinical trial 臨床試験
enroll 参加する，被験者となる

take into account 考慮に入れる

Defining cancer not only by its body part but also its genes is prying treatment options loose from old restrictions. A drug conventionally used for one cancer may turn out to work in another driven by the same abnormality. When the drugs trastuzumab and lapatinib, approved for breast cancer harboring a *HER2* mutation, were given to a group of patients with late-stage colorectal cancer with the same mutation, for example, nearly half lived for about a year, an unusually long time. Although such connections are still rare and preliminary, they indicate that it may be time to reconsider standard definitions of cancer.

The NCI launched one of the new collaborations—called MATCH—in August 2015. This study, which expects to enroll 840 volunteers, aims to provide the data needed for doctors to prescribe drugs to more patients based on tumor genetics. DNA from up to 5,000 tumor specimens will be sequenced to find suspicious abnormalities with matching gene-targeted drugs. When the trial started, eligible patients received one of 10 gene-drug combinations; that number has now expanded to 24. Meanwhile the American Association for Cancer Research has put an initial $2 million into a two-year project called GENIE, which will collect both tumor gene profiles and medical results for many thousands of patients from seven major cancer centers in the U.S. and Europe. This registry aims to provide information that investigators can use for many purposes, including identifying more mutations that might be amenable to targeted drugs and finding markers that can help with diagnosis or staging of tumors.

These and other efforts augur well for future improvements in genetically customized care for cancer patients. At present, however, they are dogged by skepticism about how quickly they will lead to meaningful changes. In addition, the push for targeted drugs could be undermined if pharmaceutical companies shift their focus to

Vocabulary

pry 苦労して〜する
conventionally 伝統的に

lapatinib ラパチニブ

preliminary 予備的, 暫定的

prescribe 処方する

specimen 組織試料, 標本

eligible 適した, 適格な

American Association for Cancer Research 米国がん学会

registry 登録, 記録

amenable 影響を受けやすい
stage 進行段階を決める

augur well for 〜にとってよい前兆である

dog つきまとう, 悩ませる

other up-and-coming approaches, such as immunotherapy. For now the gulf between the promise of precision medicine and the reality remains frustratingly large.

Vocabulary

up-and-coming 新進の
immunotherapy 免疫療法
gulf 溝

1995年に臨床応用されて以来，がん遺伝子の検査はずいぶん進歩した。当時調べられていたのは *BRCA1* と *BRCA2* という2つの遺伝子の変異で，治療方針を決めるのに重要な役割を果たした。これら変異のうち1つを持ち，乳がんや卵巣がんの家族歴のある女性は，自らも発病するリスクが高い。現在，そうした女性のなかには発病前に乳房や卵巣を切除することを選ぶ人までいる。

多くのがんが主に遺伝子の異常によって生じることが認識され，がん遺伝子解析は様々な悪性腫瘍において標準的な検査になっている。検査結果は治療法を決める参考になり，最も一般的な一部の変異については悪影響を妨げる薬もある。

だが，そうした薬のない変異を腫瘍組織に抱えている患者が多い。さらに，多くの変異が腫瘍の成長に果たしている役割はまだほとんどわかっていない。様々なタイプのがんについてDNA解析を行った結果，実に多様な変異が見つかったものの，この百科事典的な情報のうち治療方針の決定に役立っているものはごくわずかだ。治療に影響しうる特定の変異を調べる検査として米食品医薬品局（FDA）がこれまでに認可した方法はたったの29しかない。

この状況を変えるべく，大規模な共同研究がいくつか動き出した。医薬品の標的となる遺伝子変異を数多く特定し，より多くの患者について標的医薬を選べるようにする。2016年にはオバマ大統領が「米国がん撲滅ムーンショット」計画を発表した。遺伝子変異の特定を含め，10億ドルを投じる。だが，がん遺伝子探索研究は大規模かつ複雑になるため，当面は知識の蓄積と患者の利益の間の溝はむしろ広がり，期待される革新が実現するのは先になるだろう。米国立がん

研究所（NCI）で腫瘍疫学・遺伝学部門を率いるチャノック（Stephen Chanock）は「現在は過渡期にある」という。

ドライバーとパッセンジャー

がんを引き起こす遺伝子変異は大きく分けて2種類ある。1つは親から受け継いだ「生殖細胞系列変異」。もう1つは「体細胞系列変異」で，加齢や喫煙，環境からの他の影響などを受けて生後に生じた変異だ。生殖細胞系列変異は一部の小児がんなど進行性の腫瘍につながるものの，変異自体は比較的まれだ。大部分のがんは体細胞系列変異から生じる。

ほとんどの体細胞変異は結果的には無害で，多くが人体の品質管理機構によって修復される。だが一部は大混乱を引き起こし，細胞がとめどなく分裂・増殖するようになる。がんでは有害な変異遺伝子によって，過剰な細胞分裂を促進するタンパク質や，細胞増殖を抑制する本来の機能を失ったタンパク質が作られる。

腫瘍の増殖と存続に不可欠な変異は「ドライバー変異」と呼ばれ，その他の変異は単にその尻馬に乗っているようなので「パッセンジャー変異」と呼ばれている。各種のがんが生じるのにいくつのドライバー変異が必要なのかはわかっていない。ある研究によると平均的な腫瘍には少なくとも2つ，多ければ8つのドライバー変異が必要だが，別の研究によると20ものドライバー変異を含む腫瘍が多いという。

初期の成功

どんな腫瘍にどの遺伝子変異が重要なのかを解明するのは難しいが，1990年代後半には特定の変異を標的にした治療法が前進し始めていた。そうした初期の治療薬にメシル酸イマチニブ（商品名グリベック）とトラスツズマブ（商品名ハーセプチン）がある。前者は慢性骨髄性白血病に広く見られるドライバー変異を阻害し，後者は乳がんの約1/4の原因になっている*HER2*遺伝子の変異に対処する。その後，さらに他の標的医薬が続いた。

3年ほど前から，肺がんの患者は *ALK* という遺伝子の異常の有無を検査するのが当たり前になっている。肺がん患者の7％までが，*ALK* と別の遺伝子が融合する異常によって，腫瘍の成長を起こす異常なタンパク質が生じている。この変異タンパク質を阻害する薬があり，通常は標準的な抗がん剤よりも効果が大きい。ただ，腫瘍組織の *ALK* 遺伝子が正常な患者にはまったく効果がない。

皮膚がんの一種である黒色腫の患者にも遺伝子検査は役立つ。黒色腫患者の約半数は *BRAF* という遺伝子に変異があり，これが身体の他の部位への転移に関係している。FDA は 2011 年，この変異 BRAF タンパク質を阻害する初の薬を認可した。最近の研究によると，この薬に反応した転移性黒色腫患者の 80％近くが平均で 2 年延命した。黒色腫が広がった患者の余命 5.3 カ月と比べ，かなりの延命だ。

特定の変異を知って，ある種の薬の処方を避けることができる場合もある。例えば *KRAS* または *NRAS* という遺伝子に変異がある結腸直腸がん患者の場合，いくつかの薬は効かなくなる。

だが，がん遺伝子検査が進歩するにはいくつかの課題がある。まず，遺伝子の異常を発見するだけでは不十分だ。その異常が，がんの成長と存続に不可欠でなくてはならない。また，変異を見つける信頼性の高い検査法と，その変異を利用してがんを治療する方法が存在する必要がある。これらの要請は非常に難しい注文であることがわかった。どの変異ががんを引き起こしているのかを解明する困難に加え，後になって作用する変異を知る必要もあるのだ。腫瘍の成長につれ，新たな変異が出現する場合がある。一群の変異のうちどれがドライバーでどれがパッセンジャーなのかをその都度見分け，ある薬が効かなくなったら新たに遺伝子検査を行って次の薬を選べるようにしなくてはならない。

ドライバー変異を阻害する薬の開発も大仕事だ。体細胞変異遺伝子がコードしているタンパク質の多くはがん細胞の表面に存在していて，薬が届きやすい。だが，なかには細胞の内部奥深くに存在するものがある。細胞内に侵入できる低分子の薬は小さすぎて，通常はそれらの標的タンパク質に結合しない。この難題が壁となって，最も一般的なドライバー変異である *p53* や *RAS*, *MYC* といっ

た遺伝子の変異には対処法がいまだにない。

そして，体細胞変異遺伝子を標的にできる薬も，患者の生存期間をわずかに延ばせるにすぎないことが多い。1つの標的医薬で単一のドライバー変異を阻害して腫瘍を何とか縮小できたとしても，その薬に抵抗性のある細胞が1つでも残ったら，それが増殖して薬の効かない腫瘍ができてしまう。だからその種のがんは，HIV感染症と同様，複数の薬で治療する必要があるだろう。だが薬はそれぞれ費用がかかるし，副作用の危険も伴う。最適な治療戦略を見極める必要がある。

多くの体細胞変異がまれであることも，臨床応用を遅らせている。なかには患者の1%未満にしか見られないものもある。ある標的薬でその変異に対処できるかどうかを評価するには臨床試験が必要だが，変異自体がまれだと，十分な数の患者の参加を得て臨床試験を実施するのに長い時間を要する。

新たな方向

こうした難題を受け，研究手法と薬剤設計，個別化標的医療の推進加速を目指したインフラの開発が活発化している。これらは新たな知見を考慮に入れている。がんはこれまで，例えば乳がんとか肺がんなどというように，最初に発生した身体部位によって定義されてきたが，あるタイプの腫瘍を引き起こすことが知られている遺伝子変異が，典型的には身体の別の部位に生じる別のがんにも関連している場合があることがわかってきた。

がんを発生場所だけでなく遺伝子変異によって定義すると，従来の制限に縛られずに治療法を選択できるようになる。これまで特定のがんに使われてきた薬が，同じ遺伝子異常によって起こる別のがんにも有効であることが判明するかもしれない。*HER2*遺伝子変異を持つ乳がん向けに認可されていたトラスツズマブとラパチニブを同じ変異を持つ後期大腸がん患者に投与した結果，半数近くに約1年という異例に長い延命が見られたのが一例だ。こうした例はまだ少数で結果も暫定的なものではあるが，がんの標準的な定義を考え直すべき時に来ているのかもしれない。

2015年8月,米国立がん研究所は「MATCH」という共同研究を立ち上げた。840人の患者に参加してもらい,より多くの患者にがん遺伝子プロファイルに基づいて薬を処方できるよう,必要なデータを集める計画だ。総計5000の腫瘍組織試料から得たDNAの配列を解読し,疑わしい異常と,それにマッチした遺伝子標的薬を見つけ出す。一方,米国がん学会(AACR)は「GENIE」という2年プロジェクトにまず200万ドルを投じ,欧米の主要な7つのがんセンターから数千人分のがん遺伝子プロファイルと治療データを集めている。この情報を研究者に提供し,標的医薬に反応しそうな変異の特定や,診断や進行段階評価に役立つマーカーの発見など,様々な目的に利用してもらう狙いだ。

これらの取り組みは,がん遺伝子に基づく個別化医療を改善していくうえで結構なことだ。だが現時点では,いつになったら具体的な成果が上がるのかという懐疑的な見方がぬぐえない。さらに,医薬品企業が免疫療法など最近注目されるようになった別のアプローチに力点を移すと,標的医薬にはマイナスになるかもしれない。いまのところ,個別化医療に対する期待と現実を隔てる溝はいら立たしいほど大きなままだ。

A Surprising Fix for Sickle Cell
鎌状赤血球症に驚きの治療法

別の遺伝子変異を導入して発病を抑える

K. ワイントラウブ（サイエンスライター）

掲載：SCIENTIFIC AMERICAN May 2016, 日経サイエンス 2017 年 3 月号

Ceniya Harris, age nine, of Boston should be a very sick little girl. Both her parents unknowingly passed her a copy of the genetic mutation for sickle-cell disease, a debilitating and sometimes fatal blood disorder. With a double dose of the mutant gene, Ceniya's body produces a defective kind of hemoglobin—the molecule in red blood cells that takes oxygen from the lungs and releases it into tissues throughout the body. The flawed hemoglobin molecules can deform the normally round blood cells into a crescent, or sickle, shape, leading the cells to clump together and hinder oxygen's passage into tissues. The subsequent physiological havoc, known as a sickle-cell crisis, is incredibly painful and frequently requires emergency treatment to prevent life-threatening strokes and organ failure.

And yet the bouncy fourth grader, who is fond of sparkly shoes, is able to dance, participate in gymnastics and attend school without a hint of any health troubles. The secret to Ceniya's good fortune lies in a second genetic mutation she inherited—one that limits the aberrant curving of red blood cells. This unusual combination of genetic alterations means that she has yet to suffer a sickle-cell crisis, and her doctors believe that she will probably be protected from the effects of the defective hemoglobin for the rest of her life.

Vocabulary

mutation 変異
sickle-cell disease 鎌状赤血球症
debilitating 衰弱させる
mutant gene 変異遺伝子
hemoglobin ヘモグロビン
red blood cell 赤血球

crescent 三日月形の
sickle 鎌状の
hinder 妨げる
havoc 大混乱, 大打撃

stroke 発作
organ failure 臓器不全

bouncy 元気のよい

aberrant 異常な

defective 欠陥のある

鎌状赤血球症に驚きの治療法

For decades physicians have known that a few children like Ceniya have unusual genetic mutations that counteract the effects of the sickle-cell flaw. Researchers would like to re-create their uncommon physiology in everyone with sickle-cell anemia. Though not technically a cure, the compensatory treatment would spare many of the 300,000 infants around the world who are born every year with sickle cell and who often do not live beyond childhood. It would also make life a lot easier for the more than 70,000 individuals living with the disease in the U.S., who, despite treatment that mitigates some of the most serious effects of the condition, often die in their 40s.

Vocabulary

counteract 打ち消す, 妨げる
anemia 貧血
compensatory 補償的な

mitigate 緩和する

Investigators are now beginning to test such approaches, which depend on the precise alteration, or editing, of certain genes using new techniques in genetic engineering. (As will be addressed shortly, providing the compensatory mechanism should be easier to achieve than fixing the original sickle-cell genetic defect.)

"Gene-editing approaches have the potential to be game changing and really revolutionize the therapy," says Lloyd Klickstein of health care company Novartis, which is among the firms and universities exploring new sickle-cell treatments.

gene-editing ゲノム編集

WHAT DOESN'T KILL YOU

For a condition that causes life-threatening problems in childhood, sickle cell is surprisingly widespread. After all, if a mutation tends to kill people in childhood, you would expect few of the affected individuals to live long enough to mate and pass the trait to the next generation. The spread could be explained, however, if inheritance of just one copy of the mutation somehow protected people against a different threat to survival. In the late 1940s British scientist J.B.S. Haldane noted that inherited hemoglobin disorders are common in tropical areas where malaria is prevalent. He hypothesized that children born

trait 形質
inheritance 受け継ぐこと

prevalent 流行している
hypothesize 仮説を立てる

179

with a single mutated hemoglobin gene (which does not cause major problems) were somehow better able than their peers to fight off malaria and would survive to deliver the gene to their future children.

Subsequent studies supported Haldane's hypothesis, at least in part. Although individuals who carry a single sickle-cell gene actually can acquire malaria, they are less likely to die from the parasitic infection than those who do not have the mutation. How the altered hemoglobin confers this biological protection is still not entirely understood.

In contrast, researchers understand pretty well why two copies of the sickle-cell mutation can prove lethal. A molecule of hemoglobin is made up of four subunits: most commonly two identical proteins called alpha-globins and another pair of proteins known as beta-globins. Each of these subunits contains an iron-bearing structure, which, under normal circumstances, can grab on to or release a molecule of oxygen. Thus, each hemoglobin can carry up to four oxygen molecules. Individuals who inherit a single sickle-cell mutation produce one defective and one normal beta-globin; those who inherit sickle-cell genes from both parents produce only defective beta-globins.

When there is not enough oxygen around, these two defective beta-globins attach to each other. The connection is so tight that it causes the rest of the hemoglobin molecule to link up with other similarly affected hemoglobin molecules. The molecules end up forming long strands that distort the red blood cells into the sickle shape responsible for sickle-cell crises. Eventually the malformed molecules poke through the red blood cells in which they are found, like nails inside a plastic bag, says Matthew Heeney of the Dana-Farber/Boston Children's Cancer and Blood Disorders Center. The defective hemo-

Vocabulary

parasitic 寄生虫性の
infection 感染症

confer 付与する

alpha-globin αグロビン
beta-globin βグロビン

grab 捕捉する, つかむ

distort 歪ませる

poke through 突き破る

globin can puncture the cell, reducing its life span from the typical 120 days to fewer than 20 days. The body tries to replace these lost red blood cells, but if it cannot keep up, the resulting anemia causes its tissues to become oxygen-starved. The resulting tissue damage also triggers inflammation that can damage vessels and tissues.

LESSONS FROM NATURE

The only known cure for sickle-cell disease is bone marrow transplantation, which, in effect, provides a new circulatory system. But transplantation is expensive and requires a level of medical expertise that is unavailable in all but the wealthiest countries. Even there it is an option only for people who have an unaffected sibling with the right "tissue match." As if that were not challenging enough, the procedure itself carries about a 5 to 10 percent risk of death, presenting parents with a terrible choice between risking their child's life and relieving his or her pain.

There is another situation, however, in which everyone with sickle-cell disease gets a temporary reprieve from its life-threatening effects: during development in the womb.

A fetus has a distinct kind of hemoglobin that binds very tightly to oxygen, allowing it to compete successfully with its mother's hemoglobin for oxygen in the placenta. Sometime early in an infant's first year after birth, the production of this fetal hemoglobin usually drops off, decreasing the amount of oxygen found in red blood cells. In a child who inherits the sickle-cell flaw from each parent, cells usually start to sickle several months after birth. And the march of symptoms begins.

Intriguingly, no one ever shuts off production of fetal hemoglobin entirely. Most adults—whether they have sickle-cell disease or not—produce about 1 percent fetal

Vocabulary

inflammation 炎症

bone marrow 骨髄

sibling 兄弟姉妹

reprieve 一時的軽減, 猶予

fetus 胎児

placenta 胎盤

fetal hemoglobin 胎児ヘモグロビン

hemoglobin. In Ceniya's case, the gene that codes for fetal hemoglobin never received the message that it was no longer needed. Her hemoglobin remains 20 percent fetal, high enough to continue protecting her. The abundant oxygen it supplies to her red blood cells keeps her defective hemoglobin from behaving badly.

MAKING REPAIRS

The new idea for therapy would reawaken the fetal hemoglobin gene by disabling yet another gene—the one that essentially tells the fetal hemoglobin gene to stop working. How could giving sickle-cell patients the equivalent of a second genetic mutation be more practical than repairing the mutation that causes their condition in the first place? Because, at the moment, turning genes off is a lot easier than replacing the single mistake in the DNA molecule that causes the disease.

After decades spent studying the genetic underpinnings of sickle cell, Stuart Orkin, another researcher at Dana-Farber/ Boston Children's, recently located the precise spot in the DNA of long-lived blood-making cells (stem cells) where a tiny snip would allow for indefinite manufacture of fetal hemoglobin. By inducing this mutation, Orkin and his colleagues have found a way to make two wrongs into a right. Sangamo BioSciences, in collaboration with Biogen, is gearing up to test a gene-editing technology using so-called zinc-finger scissors to make the snip that Orkin recommends. And Novartis's Klickstein says that, among other approaches, his company hopes to

Vocabulary

reawaken 再び目覚めさせる
disable 働かないようにする

underpinning 基礎, 背景

stem cell 幹細胞
snip 切れ込み, 簡単な改変

gear up to 〜の準備をする

zinc-finger scissor ジンクフィンガーのはさみ (ジンクフィンガーヌクレアーゼ)

Technical Terms　クリスパー/キャス9（**CRISPR/Cas9**）　狙いの遺伝子を自由に改変する「ゲノム編集」技術の1つ。CRISPRは細菌のDNAに見つかったある種の反復配列のことで, 細菌の中ではこれがCasというタンパク質(酵素)とともに働いてDNAを切断している。ゲノム編集に使う場合は, 改変したい配列に結合するRNA分子を作り, それをガイド役としてCas9というタンパク質を標的部分に作用させてDNAを編集する。ゲノム編集技術にはこのほかに, 本文も言及しているジンクフィンガーヌクレアーゼという酵素を使う方法などがあるが, クリスパー/キャス9はそうした在来法よりも簡単で安価に実施できるため最も有望視され, 応用が急速に進んでいる。

do the same with a different technique, known as CRISPR/Cas9.

Other firms are exploring the possibility that protective genes can be safely added after all. One such company, known as bluebird bio (based in Cambridge, Mass.), relies on a virus to deliver an antisickling gene that triggers production of healthy hemoglobin in blood-making stem cells. A year after a French 13-year-old with severe disease received the treatment, he was faring well, with no sickling events and no need for painkillers, says bluebird bio's chief medical officer David Davidson. The group has begun a 20-person study in the U.S. to explore the procedure further.

Both these paths are fraught with dangers, however. Patients need to undergo chemotherapy to wipe out most of the existing stem cells that make the wrong kind of hemoglobin so that there is room for new cells that make the right kind. Even though such treatment would be available to more people than are candidates for bone marrow transplants, it is so toxic in its own right that it can cause cancers to develop years after treatment. And virtually all patients will become infertile as a result. Again, a terrible decision for a parent to make on behalf of a child.

Global Blood Therapeutics in South San Francisco hopes to develop a drug that achieves the same benefits as gene therapy but without the side effects. The goal is to keep mutated hemoglobins from glomming on to one another—something they cannot do if they are holding on to an oxygen molecule. So the company is developing a pill, currently called gbt440, that binds oxygen to the alpha-globins longer than usual. Delaying their release of oxygen even just a bit prevents the beta-globins from getting close enough to connect—especially in the tiny capillaries where sickling occurs most often. Global Blood

Vocabulary

CRISPR/Cas9
▶ Technical Terms

blood-making stem cell 造血幹細胞

painkiller 鎮痛剤

fraught with 〜をはらむ
chemotherapy 化学療法

in its own right それ自体で

infertile 不妊の, 生殖能力のない

glom on to 凝集する

capillary 毛細血管

Therapeutics expects to begin a clinical trial later this year.

Vocabulary

But as much as these therapies may make a difference for patients with sickle cell who live in wealthy countries, an entirely different solution is needed to help children elsewhere. "It's the undeveloped world that has the burden [of this disease]," says David G. Nathan, president emeritus of the Dana-Farber Cancer Institute and a leading sickle-cell researcher. For patients in the U.S., living with sickle cell is difficult, painful and terrifying; in the developing world, Nathan says, "it's a disaster."

ボストンに住む9歳のセニヤ・ハリス（Ceniya Harris）は，本来なら最悪の体調だったはずだ。鎌状赤血球症という病気をもたらす変異遺伝子を，両親から1つずつ，合わせて2つ受け継いだため，セニヤちゃんの身体は欠陥タイプのヘモグロビンを作り出す。ヘモグロビンは赤血球にあって，肺で酸素を取り入れ，全身の組織でこれを放出している分子だ。欠陥タイプのヘモグロビンは通常は丸い赤血球を三日月形（鎌状）に変え，赤血球どうしが凝集するため，組織に酸素が行かなくなる。この結果，ひどい痛みが生じ，命にかかわる発作と臓器不全を防ぐために緊急治療を頻繁に要する。

だが，この快活な小学4年生は，お気に入りのキラキラのダンスシューズをはいて踊ることができ，何の問題もなく元気に学校生活を送っている。その秘密は彼女が受け継いだ2番目の遺伝子変異にある。赤血球の異常な変形を抑える変異だ。この異例の変異の組み合わせにより，彼女はヘモグロビンが異常であるにもかかわらず，鎌状赤血球症の悪影響からおそらく一生を通じて守られる。

鎌状赤血球の欠陥を打ち消す特異な遺伝子変異を持つ子供がいることは数十年前から知られ，研究者たちはこの異例の効果をすべての患者に生み出そうとしてきた。そうした補償的な治療は厳密には治癒とはいえないものの，鎌状赤血球をもって生まれる世界で毎年30万人にのぼる新生児の多くを救えるだろう（現在は子供のうちに亡くなる例が多い）。また，米国に7万人以上いる患者の

生活をずっと楽にできるはずだ。患者は深刻な症状を緩和する治療を受けているにもかかわらず，40代で死亡することが多い。

　そうした補償的治療法の試験が始まりつつある。ゲノム編集という新技術を用いて，別の遺伝子を正確に変更・編集する（後述するように，鎌状赤血球症の原因遺伝子を修復するよりも補償機構を生むように遺伝子を編集するほうが容易だ）。

　「ゲノム編集が切り札となって鎌状赤血球症の治療法が革新する可能性がある」と，ノバルティスのクリックスタイン（Lloyd Klickstein）はいう。同社を含めいくつかの企業と大学が，この可能性を追求している。

変異遺伝子が2コピーあると発病

　子供の命を脅かす疾患として，鎌状赤血球症は驚くほど例が多い。しかし，この変異を持つ人が子供のうちに亡くなる場合が多いのなら，結婚して子供をもうける人は少なく，変異が次世代に継承されなくなるだろう。にもかかわらずこの形質が広く見られる理由は，変異遺伝子を1コピーだけ受け継いだ場合には，なぜか別の病気に対して強くなるからだと説明される。1940年代後半，英国の科学者ホールデン（J.B.S. Haldane）は，マラリアが流行している熱帯地域に遺伝性のヘモグロビン疾患が多く見られることに気づいた。彼は，変異ヘモグロビン遺伝子を片方の親から1コピー受け継いだ子供（健康上大きな問題は生じない）はマラリアを退ける力が通常よりもなぜか強まり，長生きしてこの変異遺伝子を子供に伝えるのだろうと提唱した。

　この仮説はその後の研究によって少なくとも部分的に裏づけられた。鎌状赤血球症遺伝子を1コピー持つ人は変異を持たない人に比べ，マラリアにかかっても死亡する例が少ないのだ。変異ヘモグロビンがこの利点をもたらす理由は，まだ完全にはわかっていない。

　一方，変異遺伝子が2コピーあると致命的になる理由はよくわかっている。ヘモグロビン分子は4つのサブユニットからなる。通常はαグロビンというタンパク質が2個と，別種のβグロビンが2個だ。これらはそれぞれ鉄原子を

含む構造を備え、その構造が酸素分子を捕捉・放出する。つまり、ヘモグロビン1分子で4個の酸素分子を運搬できる。しかし鎌状赤血球変異遺伝子を1コピー受け継いだ人は、βグロビンの片方に欠陥が生じる。変異を両親から受け継ぐと、どちらのβグロビンも欠陥タイプとなる。

酸素が十分にない場合、これら2つの欠陥βグロビンが互いにくっつく。この結合は非常に強く、ヘモグロビン分子の残りの部分が他の変異ヘモグロビン分子とつながる。こうして長い鎖ができ、赤血球が鎌状に歪む。ついにはこれらの奇形分子がポリ袋に入れた釘のように赤血球を突き破るのだと、ボストン小児病院がん・血液疾患センターのヒーニー（Matthew Heeney）はいう。このため通常は120日ほどの赤血球の寿命が20日足らずになる。人体はこれを補給すべく赤血球を作り出すが、それでも追いつかない場合には貧血によって組織の酸素が欠乏する。こうして組織に生じた損傷は炎症を引き起こし、血管や組織をさらに害する。

「胎児ヘモグロビン」の力

鎌状赤血球症を根治する唯一の既知の方法は骨髄移植、つまり血液細胞を実質的に一新することだ。だがこの治療は高額なうえ、富裕国にしかない高度な医療を必要とする。富裕国でも、患者と組織適合性の一致した兄弟姉妹がいる人に限られる。さらに、骨髄移植という処置自体が5〜10％の死亡リスクを伴うため、患者の両親はわが子の苦痛軽減と生命を危険にさらすこととの間で苦渋の選択を迫られる。

だが、鎌状赤血球症患者の誰もが一時的に致命的影響を免れている時期がある。胎児として母親の子宮内にいる間だ。

胎児は酸素と非常に強く結びつく特別なヘモグロビンを持っているおかげで、胎盤中の酸素を母親のヘモグロビンに負けずに取得できている。この「胎児ヘモグロビン」の生産は通常、生後1年以内の早い段階で低下し、赤血球中の酸素量が減る。すると、両親から鎌状赤血球変異遺伝子を受け継いだ子では赤血球が変形し始め、症状が進んでいく。

だが面白いことに，胎児ヘモグロビンの生産が完全にゼロになる人はいない。ほとんどの成人は，鎌状赤血球症患者もそうでない人も，ヘモグロビンの約1％が胎児タイプだ。セニヤちゃんの場合，胎児ヘモグロビンをコードしている遺伝子が「もう必要ないよ」との停止メッセージを受け取らなかったようで，ヘモグロビンの20％が胎児型のままだ。これは彼女を病気から守るのに十分な水準で，胎児ヘモグロビンが赤血球に酸素をたっぷり供給しているおかげで，欠陥ヘモグロビンが互いにくっつかずにすんでいる。

遺伝子編集で発症回避

新たな治療法は，この胎児ヘモグロビン遺伝子を再び目覚めさせようというものだ。この遺伝子に活動をやめるよう実質的に指示している別の遺伝子を阻害する。病気を引き起こしている変異を正すのではなく，患者に別の変異を新たに導入するのはなぜか？　現時点では，病気の原因となっているDNAの単一のミスを修正するよりも，別の遺伝子のスイッチを切るほうがずっと簡単だからだ。

やはりボストン小児病院がん・血液疾患センターにいるオーキン（Stuart Orkin）は鎌状赤血球症の遺伝的背景を長年研究し，長寿命の造血細胞（血液幹細胞）のDNA中に，小さな改変を導入すると胎児ヘモグロビンがずっと作り続けられるようになる場所を先ごろ正確に特定した。サンガモ・バイオサイエンシズはバイオジェンと共同で，ジンクフィンガーヌクレアーゼという制限酵素を用いた遺伝子編集技術によってオーキンが推奨する一塩基多型を作り出す試験を準備している。そしてノバルティスのクリックスタインは同じことをCRISPR/Cas9という別の遺伝子編集手法で実現したいと考えている。

また別の企業は，保護的な遺伝子を安全に追加する可能性を探っている。例えばブルーバード・バイオ（マサチューセッツ州ケンブリッジ）という企業は，健全なヘモグロビンの生産を引き起こす遺伝子をウイルスによって造血幹細胞に導入している。重い鎌状赤血球症を患う13歳のフランス人少年にこの治療を試したところ，1年後も症状なしに元気で，鎮痛剤の必要もないと同社の最高医学責任者ダビッドソン（David Davidson）はいう。同社はこの治療法をさらに進めるため，20人の患者を対象にした臨床試験を米国で始めた。

だし，これらの方法はいずれも危険をはらんでいる。まず，変異ヘモグロビンを作り出している既存の血液幹細胞を化学療法で一掃し，正しいヘモグロビンを作る新細胞に場所を空けてやる必要がある。この処置が可能な患者は骨髄移植を受けられそうな患者より多いものの，化学療法自体の毒性が強いため，数年後にがんを引き起こす恐れがあるほか，実質的に全員が不妊となる。やはり両親はわが子に代わって苦しい選択を迫られる。

サウスサンフランシスコにあるグローバル・ブラッド・セラピューティクスは，遺伝子治療と同じ効果があって副作用のない薬を開発しようとしている。変異ヘモグロビンが凝集するのを阻止しようという考え方だ。ヘモグロビンが酸素と結合していれば，凝集は起こらない。そこで同社は α グロビンと酸素分子を通常より長く結合させる錠剤「gbt440」を開発中だ。α グロビンからの酸素放出が少しでも遅れれば，β グロビンどうしが接近してつながるのを阻止できる。特に赤血球の変形が起こりやすい毛細血管において，効果が大きい。近く臨床試験を始める計画。

これらの治療法は富裕国の患者に少なからぬ恩恵となる可能性があるものの，それ以外の国の子供たちを救うにはまったく別の解決策が必要になるだろう。ダナ・ファーバーがん研究所の名誉所長で鎌状赤血球症研究の第一人者であるネイサン（David G. Nathan）は「この病気に最も苦しんでいるのは途上国だ」という。米国の患者にとって鎌状赤血球を抱えて生きるのは苦痛を伴う困難な生活だが，途上国では「災厄なのだ」。

The Paradox of Precision Medicine
個別化医療の矛盾

これまでのところ効用がはっきりせず，研究推進に疑問の声も

J. インターランディ（サイエンスライター）

掲載：SCIENTIFIC AMERICAN April 2016, 日経サイエンス 2016 年 10 月号

Precision medicine sounds like an inarguably good thing. It begins with the observation that individuals vary in their genetic makeup and that their diseases and responses to medications differ as a result. It then aims to find the right drug, for the right patient, at the right time, every time. The notion certainly has its supporters among medical experts. But for every one of them, there is another who thinks that efforts to achieve precision medicine are a waste of time and money. With a multimillion-dollar government-funded precision medicine initiative currently under way, debate is intensifying over whether this approach to treating disease can truly deliver on its promise to revolutionize health care.

Ask scientists who favor precision medicine for an example of what it might accomplish, and they are likely to tell you about ivacaftor, a new drug that has eased symptoms in a small and very specific subset of patients with cystic fibrosis. The disease stems from any of several defects in the protein that regulates the passage of salt molecules into and out of cells. One such defect prevents that protein from reaching the cell surface so that it can usher salt molecules back and forth. Ivacaftor corrects for this defect, which is caused by a handful of different

Vocabulary

precision medicine 個別化医療，精密医療
inarguably 議論の余地なく
medication 薬物療法

deliver on（約束を）果たす，期待に応える

ivacaftor アイバカフトール

cystic fibrosis 嚢胞性線維症
defect 欠陥

usher 案内する，導き入れる

genetic mutations and is responsible for roughly 5 percent of all cystic fibrosis cases. Genetic testing can reveal which individuals are eligible for this treatment.

The U.S. Food and Drug Administration fast-tracked development of ivacaftor a few years ago, and the drug been hailed ever since as the very essence of what of precision medicine is all about. Indeed, when President Barack Obama announced the launch of the government-funded precision medicine initiative in January 2015, he, too, sang ivacaftor's praises: "In some patients with cystic fibrosis, this approach has reversed a disease once thought unstoppable." Later the president declared that precision medicine "gives us one of the greatest opportunities for new medical breakthroughs that we have ever seen."

But ask opponents for an example of why precision medicine is fatally flawed, and they, too, are likely to tell you about ivacaftor. The drug took decades to develop, costs $300,000 a year per patient, and is useless in the 95 percent of patients whose mutations are different from the ones that ivacaftor acts on.

Moreover, a recent study in the *New England Journal of Medicine* found that the extent to which ivacaftor helped its target patients was roughly equal to that of three far-lower-tech, universally applicable treatments: high-dose ibuprofen, aerosolized saline and the antibiotic azithromycin. "These latter innovations are part of many small-step improvements in [cystic fibrosis] management that have increased survival rates dramatically in the past two decades," says Nigel Paneth, a pediatrician and epidemiologist at Michigan State University. "They cost a fraction of what the [high-tech] drugs cost, and they work for every patient."

The same paradox applies to nearly every example of precision medicine you can find: clinicians viewed the

Vocabulary

mutation 変異
eligible 適格な

U.S. Food and Drug Administration 米食品医薬品局
fast-track 速やかに進めさせる
hail 称える，認める
what~is all about 〜の本質

sing one's praises 〜をほめちぎる

act on 作用する

applicable 適用できる
dose 投与量
ibuprofen イブプロフェン
aerosolize エアロゾル化した
antibiotic 抗生物質
azithromycin アジスロマイシン
pediatrician 小児科医
epidemiologist 疫学者

use of a patient's genotype to determine the right dose of the anticlotting medication warfarin as a godsend until some studies suggested that the approach did not work any better than dosing through old-fashioned clinical measures such as age, weight and gender. And the drug Gleevec was hailed as an emblem of targeted cancer therapy when it shrank tumors in a subset of leukemia patients with a very specific mutation in their tumors. But then in a lot of patients, tumors developed new mutations that made them resistant to the drug, and when they did, the cancer returned. Gleevec bought many of these patients time—a few months here, a year there—but it did not change the final outcome.

The debate over the merits of precision medicine has its roots in the Human Genome Project, the 13-year, $3-billion (in 1991 dollars) effort to sequence and map the full complement of human genes. Building on that work, scientists devised a shortcut for linking particular gene variants to specific diseases with as little sequencing as possible. That shortcut, known as GWAS, for genome-wide association studies, involved examining selected sites across the genome to see which ones differed consistently between individuals who suffered from a certain medical condition and individuals who did not. Hoping for a bonanza of new drug targets, pharmaceutical companies invested heavily in GWAS. But the approach proved poor at exposing the genetic roots of disease. Study after study turned up many clusters of gene variants, any one of which could predispose someone to a condition. In most cases, these variants nudged risk up or down only by a tiny

Vocabulary

genotype 遺伝子型
anticlotting medication 抗血栓薬
warfarin ワルファリン
godsend 天の賜
Gleevec グリベック
tumor 腫瘍
leukemia 白血病

resistant 耐性のある, 薬が効かない

Human Genome Project ヒトゲノム計画

shortcut 近道, 便法
gene variant 変異遺伝子
GWAS ▶ Technical Terms

bonanza 大当たり

predispose 罹患しやすくする
nudge 少し動かす

Technical Terms　ゲノムワイド関連解析(**GWAS;** genome-wide association studies)　特定の病気の原因と思われる遺伝子変異を見つけることを狙いに近年広く用いられるようになった手法で, 大勢の人のゲノムの広い範囲を解析し, 患者とそうでない人の間で一貫して異なっている配列を統計的手法を駆使して探し出す。単一あるいは少数の遺伝子変異が原因である場合は有効だが, 多数の遺伝子が関与している疾患の場合は病気の発症と弱い関連性を持つ遺伝子変異がたくさん見つかる結果となり, 実際の治療法開発には結びつきにくい。

sliver, if at all. The results cast a pall on the notion of studying genetic variation to develop targeted therapies on a large scale.

Proponents of precision medicine argue that the problem is not the notion of exploring genetic differences per se but the extremely limited scope of GWAS. Instead of looking for a few types of common gene variants that correlate with disease, they say, researchers need to examine the entire genome—all six billion nucleotides, the building blocks of DNA. And they need to superimpose those data on top of several other layers of information about everything from family history to the microbes that inhabit the body (the microbiome) and the chemical modifications to DNA that affect how active individual genes are (the epigenome). If they compared all the data, among as many individuals as possible, they would finally be able to pinpoint which constellation of forces drive which diseases, how best to identify those forces and how to devise treatments that target them.

The precision medicine initiative that President Obama announced last year aims to do exactly that. Its centerpiece is a million-person cohort, from whom data of every conceivable kind—including genome, microbiome, epigenome—will be collected and stored in one colossal database, where scientists can access it for an endless array of studies and analyses.

To understand how all these data are supposed to help scientists conquer humanity's diseases, consider the example of warfarin. Knowing how fast or how thoroughly a person is apt to metabolize the drug should have made it easier to determine the best dose for that individual and should therefore have led to better outcomes. So why didn't it? Might diet or other factors play a role? Scientists do not know, but with a million-person cohort, they think they might be able to find out. "I guarantee that there

Vocabulary

sliver 少量
cast a pall on 〜に水を差す

per se それ自体

correlate with 相関する
nucleotide ヌクレオチド

family history 家族歴
microbiome マイクロバイオーム, 微生物叢
▶ 18ページ Technical Terms
epigenome エピゲノム

constellation 一群, 型

cohort 集団, コーホート
colossal 巨大な

apt to 〜しやすい
metabolize 代謝する

would be tens of thousands of them taking [warfarin]," says Francis Collins, director of the National Institutes of Health. "With that many subjects, you'll be able to say, 'Well, actually it does look like it helps this subset, but they happened to have a diet that was this form instead of that form.'" Furthermore, he notes, one would be able to see the subtleties of why and how a treatment works or does not work.

One thing supporters and detractors of the new initiative agree on is that the challenges of such an undertaking will be mammoth. It will require integrating terabytes of existing health data, spread across scores of databases whose content and quality will vary widely. And it will involve storing blood and tissue samples from one million people—no small feat, especially if those samples are collected at regular intervals. If it succeeds—if scientists find reliable predictors of disease in that mass of data and then devise ways to treat individual patients by targeting those predictors—doctors will still need to become fluent in this new language. Most physicians are not trained to make sense of existing genetic tests, and so far no one has come up with a good way to train them.

In theory, personalized medicine could work like Netflix and Amazon. They know every book and movie you have bought in the past few years, and armed with that information, they can predict what you are likely to purchase next. If your doctors had that kind of information at their fingertips—not about your purchase history but about how you live, where you work, what your genetic predispositions are, and which microbes are populating your skin and gut—then maybe cures would finally come like movie recommendations do.

But it seems fair to say it will be a very long time before science gets to the point where it can offer individually tailored treatment to the masses, if it ever does. The question is, Should it even try? Although precision medicine might make sense for people with certain conditions that are difficult and expensive to treat, such as autoimmune diseases, critics argue that on the whole, simpler approaches to treating disease are better because they cost less and benefit far more patients. "Let's say we find a [targeted] drug that can lower risk of diabetes by two thirds," Paneth says. "It would cost about $150,000 [a year per person] for that drug if we had it. A simple program focused on diet and exercise will do the same. Life span has increased by about a decade in the past 50 years. And none of that gain is related to DNA. It's learning about smoking and diet and exercise. It's old-fashioned stuff."

In the end, this moon shot may make more sense as a research enterprise than a public health initiative. Scientists learn more every day about the distinct forces that interact to produce disease in individuals. It is natural and fitting that they should start putting that information together in a systematic way. But society should not expect such efforts to completely transform medicine any time soon.

Vocabulary

autoimmune disease 自己免疫疾患

diabetes 糖尿病

moon shot 月着陸(のような大事業)

fitting 適切な

「個別化医療（精密医療）」というと，悪いはずがないように聞こえる。この先端医療は，人はそれぞれ遺伝子構成が異なるため，病気とその治療に対する反応も異なるという観察結果に基づいている。本人の遺伝情報をもとに，患者に適切な薬を見つけて適切なときに投与するのが個別化医療の目的であり，医療専門家の一定の支持を得ている。しかし，個別化医療を実現しようという努力は時間とお金のムダであると考えている専門家が支持者と同じだけいる。米国政府が個別化医療イニシアティブに巨費を投じるなか，個別化医療が本当に医療を革新できるのかどうか，議論が白熱している。

個別化医療で何が実現できるかを支持派の科学者に尋ねると，よく例に挙がるのが，嚢胞性線維症患者のうち少数かつ特定の人の症状を緩和できる「アイバカフトール」という薬だ。この病気は，細胞の内外に塩分を輸送するイオンチャネルのタンパク質に欠陥があるために起こり，欠陥の種類は複数ある。ある欠陥はこのタンパク質が細胞膜に到達するのを妨げ，正常なチャネルが形成されずに塩分の出し入れができなくなる。アイバカフトールはこれを正すのだが，この変異を持つ人は一部で，嚢胞性線維症患者の約5%だ。患者の遺伝子を調べて，この治療にふさわしい人を判別する。

米食品医薬品局（FDA）は数年前にアイバカフトールの開発を強く促進し，以来この薬は個別化医療の本質を示す代表例とされてきた。オバマ大統領も2015年1月に個別化医療イニシアティブを発表した際に，この薬をほめちぎった。「一部の嚢胞性線維症患者について，このアプローチはかつて手の打ちようがないと考えられていた病気を逆転させた」。大統領は後に，個別化医療が「医療の新たなブレークスルーにつながるこれまでで最大の機会をもたらす」とも述べた。

だが，個別化医療が致命的欠陥を抱えている理由を批判派に尋ねると，よく例に挙げられるのが，やはりこのアイバカフトールだ。この薬は開発に数十年を要したほか，患者1人の薬代が年間30万ドル（約3000万円）にもなり，アイバカフトールが作用するのとは異なる変異を持つ95%の患者にはまるで役立たない。

さらに，New England Journal of Medicine 誌に掲載された最近の研究によると，アイバカフトールが適応患者の病状を改善する程度は，はるかにローテクで患者全員に適用できる3つの治療法とほぼ同じだ。①高用量のイブプロフェン②食塩水エアロゾル吸入③抗生物質のアジスロマイシン――の3つで，「これらは嚢胞性線維症治療における数多くの小さな改善の一部であり，これらによって患者の生存率は過去20年で劇的に向上した」とミシガン州立大学の小児科医・疫学研究者パネート（Nigel Paneth）はいう。「費用はハイテク医薬のほんの一部ですみ，しかも患者全員に効く」。

これと同じ矛盾が個別化医療のほぼすべての例に見られる。抗血栓薬ワルファリンの適切な投与量を決めるのに患者の遺伝子型を利用できることがわかり，臨床医はそれが最善の方法だと考えていたが，後に年齢・体重・性別など昔ながらの臨床基準で判断するのと変わらないことが複数の研究で示唆された。抗がん剤グリベックは腫瘍細胞に特定の変異が見られる一部の白血病を改善して，分子標的医薬の象徴としてもてはやされた。しかしその後，多くの患者では腫瘍細胞が新たに変異してグリベックに耐性を獲得し，がんが再発した。グリベックは多くの患者の命を延ばしたが（2〜3カ月，あるいは1年），結局は救命に至らなかった。

個別化医療の利点をめぐる論争の原点は「ヒトゲノム計画」にある。13年の期間と30億ドル（1991年のドル価）をかけてヒトゲノム配列をすべて解読したプロジェクトだ。科学者たちはこの成果を土台に，最小限の配列解読によって特定の遺伝子変異を特定の病気に結びつける便法を編み出した。「ゲノムワイド関連解析（GWAS）」と呼ばれる方法で，ゲノムからいくつかの場所を選んで，ある病気の患者と健常者で違いが一貫して見られる場所を探す。製薬各社は新薬の標的となる大当たり発見を期待して，ゲノムワイド関連解析に重点投資した。だが，病気の遺伝的基盤を明らかにする方法としては結局のところお粗末だった。解析のたびに，病気への罹患しやすさと関連しうる遺伝子変異群が次々に現れた。ほとんどのケースでは，これらの変異はリスクを左右するとしてもほんのわずかだった。この結果は，遺伝子変異を調べて標的医薬を大量に開発するという考え方に水を差した。

こ れに対し個別化医療の推進者は，遺伝的差異を探すという考え方そのものには問題はなく，ゲノムワイド関連解析の調査範囲が限られているのが問題なのだと反論する。病気と相関する少数タイプの一般的な変異遺伝子を探すのではなく，ゲノム全体，つまりDNAを構成する60億の塩基を調べる必要がある。そしてそのデータを，家族歴から本人のマイクロバイオーム（人体にすんでいる微生物），エピゲノム（個々の遺伝子の発現に影響するDNAの化学修飾）まで，あらゆる階層の情報に重ね合わせて解析する必要がある。できるだけ多くの人についてすべてのデータを比較すれば，どの要素の組み合わせがどの病気をもたらしているのかをついに特定できるはずだ。それらの病因を突き止める最良の方法と，その標的に応じた治療法の開発も見えてくるだろう。

 オ バマ大統領が昨年に打ち出した個別化医療イニシアティブは，まさにそれが狙いだ。100万人の大集団について，ゲノムやマイクロバイオーム，エピゲノムなど考えうるあらゆるデータを収集して巨大データベースに蓄積し，科学者がこれにアクセスして様々な研究と解析を続けられるようにする。

 こ れらのデータがいかに役立つか，先に触れたワルファリンの例を考えてみよう。患者がこの抗血栓薬をどれだけ素早く代謝する体質であるかがわかれば，最適な投与量を決めやすくなり，よい結果につながるはずだ。なのに，そうならなかったのはなぜだろうか？　食事など他の要因が関係しているのだろうか？　現在は不明だが，100万人の集団があれば解明できる可能性がある。「この集団は数万人のワルファリン服用者を含むことになるに違いない」と米国立衛生研究所（NIH）所長のコリンズ（Francis Collins）はいう。「それだけいれば，『おや，このグループでは確かにうまくいっているようだが，食事内容に他とは違う傾向があるな。そのせいかもしれない』などといえるようになるだろう」。さらに，薬の効き目が人によって異なる微妙な理由も見えてくるだろうという。

 新 イニシアティブの支持者と反対者が一致しているのは，それがとてつもない大事業になるだろうという点だ。内容と質が様々な多数のデータベースに分散して存在する膨大な健康データを統合する必要がある。また，100万人から血液試料と組織試料を採取して保存することになるだろう。これは大仕事で，定期的に収集するとなるとなおさらだ。これがうまくいって，信頼性の高い病気

4　先端医療

予測因子が見つかり，それに基づく治療法を開発できた場合も，一般の医師がそうした新手法に習熟する必要があるだろう。既存の遺伝子検査データを解釈する訓練を受けた医師はまだほとんどいないし，よい訓練法も考案されていない。

　理屈のうえでは，個別化医療はネットフリックスやアマゾンのように機能するだろう。これらのネット企業は顧客がここ数年に購入した本と映画配信を知っており，その情報をもとに，顧客が次に購入しそうなものを予測できる。もしあなたの主治医がこうした情報を自由に利用できたなら（購入履歴ではなく，あなたの生活と仕事，遺伝的な疾病素質，皮膚と腸のマイクロバイオームなどの情報を利用できたら），ネット企業がお薦め映画のタイトルを表示するように，簡単に"お薦め治療法"を割り出せるかもしれない。

　だが，科学が個別化医療を大勢の人々に提供できる日がくるとしても，それはずっと先になるとみるべきだろう。問題は「そもそもそれを目指すべきなのか」だ。自己免疫疾患など治療が困難で治療費もかさむ特定の病気については，個別化医療は意義があるかもしれないが，総じていえば，もっと単純な治療法のほうがよいと批判派は主張する。費用がかからず，より多くの患者に益があるからだ。「糖尿病のリスクを2/3に下げられる標的医薬が見つかったとしよう」とパネートはいう。「ただ，薬代は患者1人で年間15万ドルになるという。食事と運動に絞った単純な健康プログラムで同じ効果が得られるだろう。過去50年で平均寿命は10年ほど延びた。DNA解析とは無関係だ。喫煙と食事，運動について人々が知ったことによる。ハイテクではない」。

　つまるところ，個別化医療という大事業は公衆衛生というよりは科学研究としての意味が大きいのだろう。相互作用を通じて個人に病気を引き起こす個別要因について，科学者は毎日理解を深めている。その情報を体系的に組み合わせる試みを始めるべきだというのは，自然で適切な考え方だ。だが，そうした取り組みが早々に医療を変革すると期待すべきではないだろう。

The Not So Silent Epidemic
広がる睡眠時無呼吸症

いびきは危険な無呼吸症のしるしかもしれない
神経刺激の新治療法などが試みられている

D. ヌーナン（サイエンスライター）

掲載：SCIENTIFIC AMERICAN June 2015, 日経サイエンス 2015 年 12 月号

Every night, before he goes to sleep, Al Pierce, whose thunderous snoring used to drive his wife out of their bedroom, uses a small remote control to turn on an electronic sensor implanted in his chest. The sensor detects small changes in his breathing pattern—early signs that Pierce's airway is beginning to collapse on itself. When the device senses these changes, it triggers a mild jolt of electricity that travels through a wire going up his neck. The wire ends at a tiny electrode wrapped around a nerve that controls muscles in his tongue. The nerve, stimulated by the charge, activates muscles that thrust Pierce's tongue forward in his mouth, which pulls his airway open.

Throughout the night the 65-year-old plumber in Florence, S.C., gets hundreds of little jolts, yet he sleeps quietly. In the morning, rested and refreshed, Pierce uses the remote to turn off the device.

This new technology, called upper-airway electronic stimulation and approved by the U.S. Food and Drug Administration last summer, offers much more than relief from an annoying noise. Pierce's loud snoring was the most obvious symptom of obstructive sleep apnea, a drastically underdiagnosed disorder shared by an estimated 25 million Americans. It can lead to high blood pres-

Vocabulary

snoring いびき
drive out of 〜から追い出す

collapse つぶれる, 虚脱する
jolt 瞬間的な刺激

thrust 押し出す

plumber 配管工

upper-airway electronic stimulation 上気道電気刺激
U.S. Food and Drug Administration 米食品医薬品局
obstructive sleep apnea 閉塞性睡眠時無呼吸症

sure, heart disease, diabetes, depression and an impaired ability to think clearly. Overall, people with severe sleep apnea have triple the risk of death from all causes as compared with those without the disorder.

Yet help has not been easy for sufferers to find. One very effective option, a strap-on mask that gently pushes air into the throat to hold it open, is rejected by a great many of the people who try it because the device is uncomfortable. Other alternatives offer only mixed results. So a surgical implant and nerve stimulation, as extreme as it may sound, could be the answer for many. In a study published last January in the *New England Journal of Medicine*, the technique reduced episodes of severe apnea by about two thirds. The FDA approval opens the door to insurance coverage for the treatment.

Doctors, for several reasons, have not been pushing to find apnea therapies. Patients tend not to bring serious apnea up with physicians as a problem, for one thing. And doctors may have their own reasons for treating the disorder lightly. "Sleep apnea is not on a death certificate," says Patrick J. Strollo, Jr., a sleep specialist at the University of Pittsburgh Medical Center. "While it may contribute to death, it's not really a direct cause." So, he says, "there is less urgency from primary care doctors and other doctors to address this problem."

Pierce found out that he had apnea only because his wife, Gail, asked her doctor for a prescription for sleeping pills. He asked why, and Gail explained that she needed them because of her husband's snoring. About half of the people who snore loudly have sleep apnea, according to the National Sleep Foundation. The doctor told her that if things were that bad, her husband should come in for a sleep study: an overnight observation period during which various sensors are attached to a patient. The study revealed that Pierce was having as many as 30 apnea epi-

Vocabulary

diabetes 糖尿病
sleep apnea 睡眠時無呼吸症

surgical implant 外科手術で埋め込む装置

episode 症状の発現

death certificate 死亡証明書

primary care doctor かかりつけ医, 一般開業医

prescription 処方

sodes an hour. Despite years of feeling tired all the time, he was stunned that he had an actual medical problem. "I thought that was the way everyone lived. I didn't know any different," Pierce recalls.

Obstructive sleep apnea often develops when people age or put on extra weight. Fat narrows the tube of the airway, and the muscles in the mouth and throat also can lose their tone. When these muscles further relax during sleep, the airway becomes constricted and blocks the flow of air to the lungs. Some people with severe apnea stop breathing altogether, for up to a minute or two, as many as 600 times a night. This oxygen deprivation forces the heart to work harder and creates surges of adrenaline, which in turn cause blood pressure to spike. In addition, fluctuating oxygen levels can cause cell and tissue damage in the lungs and other organs.

Major interventions such as reconstructive throat surgery have often been ineffective. Physicians frequently recommend lifestyle-based changes such as losing weight and sometimes even playing the didgeridoo, a large Australian wind instrument that strengthens and tones the muscles of the tongue. Nose strips and generic mouthpieces, readily available over the counter, target snoring, the symptom, rather than sleep apnea, the underlying problem. The trouble is, what helps one patient may fail another completely. Plus, anything designed to go into the mouth or throat during sleep, to prop the airway open, can bother the patient and actually disrupt sleep. Any treatment has to be comfortable, easy to use and reliable.

Difficulty meeting all those criteria is what bedevils the strap-on mask, called CPAP, for continuous positive airway pressure. The oxygen mask covers the nose (or the nose and mouth) and is held in place by straps that wrap around the head. A small bedside pump delivers a steady flow of pressurized air to the mask through plastic tubing.

Vocabulary

stun 唖然とさせる

tone 筋肉の緊張状態, 張り
constrict 狭める

deprivation 欠乏
adrenaline アドレナリン

intervention 介入処置
reconstructive throat surgery 喉頭再建手術

didgeridoo ディジュリドゥ

tone 正常にする

symptom 症状

prop 支える

bedevil 苦しめる, 悪くする
CPAP 持続陽圧呼吸療法装置, シーパップ

The therapy, available since the early 1980s, almost guarantees relief from obstructive sleep apnea symptoms, and research shows that it lower rates of cardiovascular disease and death in patients who use it.

Use is the key: fully half of the people who try the mask abandon it. Pierce is one of them. "I was miserable," he says. Like so many others, Pierce could not sleep easily while wearing something over his face, and he did not like the way the tubing restricted his movements in bed.

Strollo is a strong CPAP advocate but has long recognized the need for an alternative. Upper-airway electronic stimulation could be that option, he says. Strollo led a large study of the new treatment, a yearlong safety and efficacy trial involving 126 people with moderate to severe obstructive apnea. The participants all had a body mass index (BMI) of 32 or less (a man who is five feet, 10 inches in height and 223 pounds in weight has a BMI of 32), had tried CPAP first and had no history of cardiovascular disease. In last January's *New England Journal of Medicine* study, Strollo and his colleagues reported that the therapy, with a device made by Inspire Medical Systems, reduced subjects' sleep apnea events by 68 percent, from a median of 29.3 events an hour to nine an hour, basically turning severe apnea into a mild case. (CPAP, after adjustment, can do even better. It can cut the number of severe apnea events to fewer than five an hour, on average, but only in patients who stick with it.)

Alan R. Schwartz, a sleep specialist at Johns Hopkins University who did much of the early work on nerve stimulation—he showed in animals that jolting the tongue-controlling nerve would open their airway—says he is pleased but cautious. "We've still got a lot to learn," he notes, pointing out that overweight and obese people, who make up a significant percentage of the obstructive apnea population, are not considered good candidates for the procedure because of their excess airway tissue.

What is more, stimulation involves an invasive procedure. The surgery to implant the device takes about two hours. A head and neck surgeon, working through an incision in the side of the neck, under the patient's jaw, places an electrode on the hypoglossal nerve, which controls the muscles of the tongue. The surgeon also puts a battery pack and a sensor in the chest and connects them to the electrode with a wire lead. The patient usually can go home a day later; the device is turned on and adjusted after a month.

Researchers are investigating more alternatives, such as medication. In a six-week trial involving 120 patients, David W. Carley, a physician at the University of Illinois at Chicago, is testing a drug called dronabinol, which is a synthetic version of an active compound in marijuana. He is comparing people who get the drug with those who do not. Dronabinol may prevent or reduce sleep apnea episodes by stimulating certain neurotransmitter activity in the brain. Other researchers are looking at the role played by leptin, a hormone that suppresses appetite and may improve respiratory function. A small study of 26 obese subjects with BMIs greater than 45 suggests that certain levels of leptin may minimize upper-airway collapse.

Schwartz is also trying to modify the stimulation technique, testing a device that eliminates the sensor. Instead it sends a repeated charge to the nerve in the tongue during the night to keep the airway open. This refinement should simplify the surgery and reduce parts that could fail, Schwartz says.

Pierce, however, is quite happy with the system he has. When he is awake—or quietly sleeping—he does not even notice it.

Vocabulary

invasive 侵襲的な

incision 切開
hypoglossal nerve 舌下神経

medication 薬物療法

dronabinol ドロナビノール

neurotransmitter 神経伝達物質
▶ 72ページ Technical Terms
leptin レプチン

サウスカロライナ州フローレンスに住む65歳の配管工ピアース（Al Pierce）は，かつてそばで寝ている妻が寝室から逃げ出すほどの大きな鼾（いびき）をかいていたのだが，いまは毎晩寝る前に小さなリモコンを操り，胸部に埋め込んだセンサーのスイッチを入れる。このセンサーが呼吸リズムのわずかな変化を検知する。気道が虚脱し始めたことを示すこの初期兆候を装置が検知すると，首に埋め込まれた電線に軽い電気刺激が伝えられる。電線の終端は舌の筋肉を制御する神経に巻き付いており，電気刺激を受けてこの神経が発した指令によって舌の筋肉が舌を前方に動かし，気道を開く。

ピアースが一晩に受ける電気刺激は数百回に上るが，本人は安眠している。朝，爽やかに目覚めると，リモコンで装置のスイッチを切る。

米食品医薬品局（FDA）が2014年夏に認可したこの新技術は「上気道電気刺激」と呼ばれ，はた迷惑な鼾を抑える以上の効用がある。ピアースの大きな鼾は，「閉塞性睡眠時無呼吸症候群」の最も顕著な症状なのだ。この病気を抱える米国人は2500万人に上ると推定されるが，医師にかかってそう診断されている例はごく一部。高血圧や心臓病，糖尿病，うつ病，明晰な思考が損なわれるといった状態につながる恐れがある。重度の睡眠時無呼吸を患っている人は死亡リスクが3倍になる。

だが，患者に役立つ簡単な方策がない。非常に効果的な方法として，喉に空気を静かに送り込んで気道を開く外付け式のマスクがあるのだが，装着感が悪いので多くの患者は使おうとしない。他の方法は効果がまちまちだ。そこで，装置を外科手術で埋め込んで神経を刺激するという大げさにも思える方法が，多くの患者の解決策となりうる。2015年1月に *New England Journal of Medicine* 誌に発表された研究によると，この方法によって重大な無呼吸症状の発現が約2/3に減った。またFDAの認可を得たことで，保険適用の道も開かれた。

医師たちはいくつかの理由から，無呼吸症の治療法開発にあまり積極的ではなかった。ひとつには，重い無呼吸症の患者もそれを問題とは感じずに，医師の診察を受けない傾向がある。医師もこの疾患を軽視してきたが，それには理由がある。「睡眠時無呼吸は死亡証明書に記載するような死因ではない」と，

ピッツバーグ大学メディカルセンターの睡眠専門家ストロロ（Patrick J. Strollo, Jr.）はいう。「無呼吸症は死につながるかもしれないが，直接の死因にはならない」。だから「かかりつけ医をはじめとする医師にとって，この病気は緊急性が低いのだ」という。

ピアースは妻のゲイルが医師に睡眠薬の処方を求めるまで，自分が無呼吸症だとは知らなかった。なぜ睡眠薬がいるのか尋ねたところ，あなたの鼾のせいで必要なのよ，という答えだった。米国立睡眠財団によると，大きな鼾をかく人のほぼ半数が睡眠時無呼吸症だ。ゲイルは主治医から，そんなにひどいのならご主人は睡眠検査を受けるべきだと勧められた。患者に様々なセンサーをつけて，一晩中ずっと観察する。この検査の結果，ピアースは睡眠中に無呼吸になる事例が1時間あたり30回もあることがわかった。彼は何年も前から常に疲労を感じてはいたが，実際に医学上の問題があると知って仰天した。「疲れを感じるのはみな同じだと考えていた。自分がほかと違うとは思いもしなかった」と回想する。

閉塞性睡眠時無呼吸症候群は加齢や体重増に伴って発症することが多い。脂肪が気道を狭め，口と喉の筋肉が正常な張りを失う場合もある。睡眠中にこれらの筋肉がさらに弛緩すると，気道が狭まって肺への空気の流れが妨げられる。重度の無呼吸症患者のなかには，長い場合には1～2分間も呼吸が完全に止まり，そうした呼吸停止状態が一晩に最高で600回も生じる例がある。この酸素欠乏によって心臓はさらに激しく働く必要に迫られ，アドレナリンが大量に放出されて血圧が急上昇する。さらに，酸素レベルの変動が肺などの臓器の細胞と組織にダメージを及ぼす。

喉頭再建手術などの大手術はあまり効果がないことが多い。医師が推奨するのは減量など生活習慣の改善による方策で，舌の筋肉を強化して正常に戻すためにディジュリドゥというオーストラリアの木製の大きな管楽器の演奏を勧める医師もいる。鼻ストリップやマウスピースなど薬局ですでに入手できる用具は鼾という症状を抑えるもので，背景にある睡眠時無呼吸を治す効果はない。厄介なのは，ある患者に有効だったものが別の患者には効かない場合があることだ。さらに，睡眠中に口や喉に入れて気道を支えるように設計されたものはみな，患者にとってうっとうしく，睡眠を妨げてしまう。苦にならず，使いやすくて，信

頼性の高い治療法が必要だ。

CPAP（シーパップ；continuous positive airway pressure ＝持続陽圧呼吸療法装置）というマスクの悩みも，これらの要請にこたえられないことにある。この酸素マスクは鼻（あるいは鼻と口）にかぶせ，頭にストラップをかけて固定する。ベッドサイドに置いた小型のポンプから，ビニール管を通して常に空気を送る仕組みだ。1980年代初めに実用化し，閉塞性睡眠時無呼吸の症状をほぼ確実に緩和できる。これを利用した患者では心血管疾患の発症率と死亡率が下がることが調査研究で示されている。

使用法がカギだ。このマスクを試した患者の優に半数が使用をやめてしまう。ピアースもそうだった。「惨めな思いがした」と彼はいう。多くの試用者と同様，顔に何かを付けた状態では容易に眠れなかったし，チューブによってベッド上の動きが制限されるのも不快だった。

ストローロはCPAPを強く推奨しているものの，代替法の必要性もかねて認識してきた。上気道電気刺激は1つの選択肢になりうると彼はいう。彼は中程度ないし重度の閉塞性無呼吸症患者126人を被験者としてこの方法の安全性と効果を調べる1年間の大規模臨床試験を率いた。被験者はみな体格指数BMIが32以下（身長178cm体重100kgの人のBMIが32）で，以前にCPAPを試した経験があり，心血管疾患の既往歴のない人たちだ。ストローロらは2015年1月に*New England Journal of Medicine*誌に発表した研究で，インスパイア・メディカル・システムズ社の装置を用いた上気道電気刺激法によって被験者の睡眠時無呼吸事象が約68％減り（1時間あたり中央値で29.3回だったのが同9回に），重度の無呼吸症が基本的に軽度に緩和したことを報告した（一方，調整後のCPAPはさらによい効果を上げており，重度の無呼吸事象を平均で毎時5回未満に削減できている。ただし継続使用の場合に限る）。

舌を制御している神経を軽く電気刺激すると気道が開くことを動物実験で示すなど，神経刺激法の初期の研究の大部分を行ったジョンズ・ホプキンス大学の睡眠研究専門家シュワルツ（Alan R. Schwartz）は，この成果を歓迎しつつ「解明しなくてはならないことがまだ多く残っている」と注意を促す。閉塞性無呼

吸症患者のかなりの部分を占める体重過多ないし肥満の人たちは，気道の組織が過剰なため，この方法に向くとは考えられないと指摘する。

 さらに，電気刺激には手術が必要だ。装置を埋め込む手術は約2時間かかる。首の側面，顎の下あたりを切開し，舌の筋肉を制御している舌下神経の上に電極を設置する。またバッテリーパックとセンサーを胸部に埋め込み，リード線で電極につなぐ。患者はだいたい翌日に帰宅できるが，装置にスイッチを入れて調整するのは1カ月後になる。

 薬剤など別の治療法も研究されている。イリノイ大学シカゴ校の内科医カーレー（David W. Carley）はマリファナの活性成分を人工的に合成した「ドロナビノール」という薬を120人の被験者で試す6カ月の臨床試験を行っている。この薬を投与された患者と投与されなかった患者を比較する。ドロナビノールはある種の神経伝達物質の脳での活性を刺激することによって，睡眠時無呼吸の症状発生を防止または抑制できる可能性がある。また別の研究者は，食欲を抑制し呼吸機能を改善できるとされるレプチンというホルモンの役割に注目している。BMIが45以上の肥満者26人を対象にした小規模試験で，ある濃度のレプチンが上気道の虚脱を抑えられる可能性が示唆された。

 シュワルツは電気刺激法の改良にも取り組んでおり，センサーを省略した装置を試している。夜間に舌の神経に電気刺激を繰り返し送り続けることで，気道を開いた状態に保つ。簡単な手術ですむほか，故障の恐れがある部品も少なくなるはずだとシュワルツはいう。

 だが，ピアースは現在使っているシステムで十分満足している。目覚めている間ずっと，そして静かに眠っているときも，装置はまったく気にならないという。

原著者

ジェニーン・インターランディ（Jeneen Interlandi）　サイエンスライター
フェリス・ジャブル（Ferris Jabr）　サイエンティフィック・アメリカン編集部
ダイナ・マロン（Dina Fine Maron）　サイエンティフィック・アメリカン編集部
メリンダ・モイヤー（Melinda Wenner Moyer）　サイエンスライター
デイビッド・ヌーナン（David Noonan）　サイエンスライター
エレン・シェル（Ellen Ruppel Shell）　ジャーナリスト
デイビッド・スティップ（David Stipp）　サイエンスライター
クラウディア・ウォリス（Claudia Wallis）　フリーランスの記者・編集者
ジェシカ・ワプナー（Jessica Wapner）　サイエンスライター
カレン・ワイントラウブ（Karen Weintraub）　サイエンスライター

訳者・編者

日経サイエンス編集部

日経サイエンスで鍛える科学英語
医療・健康編

2017年12月15日　　1版1刷

編者	日経サイエンス編集部
	© Nikkei Science, 2017
発行者	鹿児島 昌樹
発行所	日経サイエンス社
	http://www.nikkei-science.com/
発売	日本経済新聞出版社
	東京都千代田区大手町1-3-7　〒100-8066
	電話 03-3270-0251（代）

印刷・製本 大日本印刷
ISBN978-4-532-52074-8

本書の内容の一部あるいは全部を無断で複写(コピー)することは、法律で認められた場合を除き、著作者および出版社の権利の侵害となりますので、その場合にはあらかじめ日経サイエンス社宛に承諾を求めてください。

Printed in Japan